Springer Series in Optical Sciences   Volume 63

Editor: Peter W. Hawkes

Springer Series in Optical Sciences

Harald Ibach

Electron Energy Loss Spectrometers

The Technology of High Performance

With 103 Figures

Springer-Verlag

Berlin Heidelberg New York London
Paris Tokyo Hong Kong Barcelona

Professor Dr. HARALD IBACH

Institut für Grenzflächenforschung und Vakuumphysik
Kernforschungsanlage Jülich GmbH, Postfach 19 13
D-5170 Jülich, Fed. Rep. of Germany

Guest Editor: Dr. PETER W. HAWKES

Laboratoire d'Optique Electronique du CNRS
Boîte Postale No. 43 47, F-31055 Toulouse Cedex, France

ISBN 3-540-52818-0 Springer-Verlag Berlin Heidelberg New York
ISBN 0-387-52818-0 Springer-Verlag New York Berlin Heidelberg

Library of Congress Cataloging-in-Publication Data. Ibach, H., 1941– Electron energy loss spectrometers: the technology of high performance / Harald Ibach. p. cm.–(Springer series in optical sciences ; v. 63) Includes bibliographical references and index. ISBN 0-387-52818-0 (U.S. : acid-free paper) 1. Electron energy loss spectroscopy. 2. Spectrometer. 3. Electron optics. I. Title. II. Series. QC454.E4I218   1990   537.5'6– dc20   90-42425

Preface

After building electron energy loss spectrometers for many years, we felt the need to develop a scientific basis for their construction. Progress in this direction was impeded by the fact that the apparently most successful designs involved cylindrical deflectors. Lens systems properly adapted to the one-dimensional focusing in cylindrical deflectors cannot possess rotational symmetry, so that standard programs could not be used. Hence we began our expedition into the well-established field of electron optics with the development of programs for lens systems that exhibit merely C_{2v} symmetry rather than rotational symmetry. The tremendous success of lenses calculated with such programs encouraged us to investigate also the other electron optical elements in a spectrometer more closely. While it has been recognized since the early work of Kuyatt and Simpson that the maximum monochromatic current that an electron optical device can produce is limited by space charge, that is, by the repulsive forces between the electrons, the details of the effect of the space charge on electrostatic deflectors were not understood. This volume is the first to describe relatively straightforward, analytical, solvable models for the effect of the space charge on the first-order focusing properties of cylindrical deflectors. The analytical considerations are then extended by numerical simulations of electrostatic deflectors under space charge conditions, and the design of the deflectors is optimised according to these considerations. Space-charge-saturated monochromators require feed beams with specified angular apertures. We have therefore devoted one chapter to the design of emission systems operating under space charge conditions at low energies.

For each of the electron optical elements necessary in electron energy loss spectrometers, we describe several possible designs, including the specification of all relevant optical and mechanical parameters. All the elements described in the book are not only analysed numerically but have also been shown to work successfully in experiments. Readers who are interested in building electron spectrometers themselves may therefore use the designs described here directly without plunging into the complexities of electron optics.

This volume could not have been written without the committed and diligent work of two collaborators, D. Bruchmann and Dr. S. Lehwald. Mr. Bruchmann performed all the engineering on the many spectrometers and variations thereof that we built, including the design of an ingenious system of digitally controlled power supplies and the operating software. Dr. S. Lehwald was an indispensable partner in ordering our thoughts and in the identification of the most relevant optical parameters at each stage of development. He also carried the burden of experimental tests on the electron optical designs, and only through the continuous interplay between experimental tests and numerical simulation was the eventual success achieved. His critical reading of the manuscript was also of tremendous help.

Jülich, July 1990   *H. Ibach*

Contents

1. Introduction

Electron energy loss spectroscopy is the name for an experimental technique in which electrons with a well-defined energy are scattered from a target and where the energy distribution, and frequently also the angular distribution, of the scattered electrons is measured. The generally used acronyms for the technique are HREELS (*H*igh *R*esolution *E*lectron *E*nergy *L*oss *S*pectroscopy) or EELS.

The mean free path of 100 eV electrons in a solid is only about 1 nm and increases slowly with increasing energy. High energy electrons are therefore required to investigate bulk properties of matter in a transmission experiment. When electrons with an energy up to a few hundred eV are backscattered from a solid the electrons interact with only the outermost atomic layers of the material. It is this surface sensitivity which has spurred the interest in and development of EELS recently, as the technique offers the possibility of probing vibrational and electronic surface excitations of solids or adsorbed molecules on surfaces. The technique has had a large impact on the development of surface chemistry and the science of catalysis [1.1]. Other recent applications include surface structure analysis [1.2] and the determination of the dispersion of surface phonons [1.3]. The physics of semiconductor surfaces [1.4] and of epitaxial growth [1.5] are further areas of recent activity. With spin polarized electron beams and spin analysis of scattered electrons even the magnetic excitation spectrum of a solid surface or of a thin film may be investigated [1.6]. In total, several hundred papers are published each year in which electron energy loss spectroscopy is used to probe vibrational or electronic properties of matter. The purpose of this volume is not to add to the various reviews in the field [1.7], but rather to focus on the instrumentation needed in EELS.

A number of technical reports on electron energy loss spectrometers have appeared over the years [1.8–13], and several improvements, including multi-channel detection [1.14] and parallel processing [1.15] have been suggested. Modern techniques of computer simulation have also been employed [1.16–18] in order to optimise one aspect or another of a spectrometer. A thorough treatment of the electron optics of spectrometers that encompasses all essential design parameters and physical requirements from the emission system to the detector has been lacking until now.

The level of presentation in this volume is such that no special knowledge of electron optics is required. The book concentrates on the electron optics specific to electron energy loss spectrometers. For a general treatise on electron optics the reader is referred in particular to the recent work of *Hawkes* and *Kasper* [1.19].

The problem of designing optimised electron energy loss spectrometers was complicated by the fact that it took many years to establish the physics of electron surface scattering so that the desired design parameters could be understood. Unlike light optics, in electron optics the space charge of the electron beam needs consideration. It has been known for some time now that the space charge in energy dispersive devices (the monochromator) is the limiting factor in producing a high monochromatic current at the sample and ultimately a high count rate in the detector [1.2, 6]. It is also experimentally established that cylindrical deflectors, used as monochromators, produce the highest monochromatic current, although the reason for this is not so obvious. The use of cylindrical deflectors raises the additional difficulty that these devices focus in only one dimension, the radial plane. It is clear that any optimised lens system for an electron spectrometer with cylindrical deflectors as the energy dispersive elements should not be rotationally symmetric around the optic axis. The calculation of the optical properties of such lenses requires trajectory calculations and an optimisation process in three dimensions, rather than in two dimensions as for rotationally symmetric lenses. In this volume we will show that such a three-dimensional optimisation, which also includes space charge, is indeed feasible without using excessively powerful computers. The final result of such optimisation will be a spectrometer design for which the count rate in inelastic scattering processes exceeds that of earlier instruments by several orders of magnitude and this will surely open the way to a new area of electron energy loss spectroscopy.

The volume is organised as follows. We begin with a general consideration of the mathematics and algorithms pertinent to the computer simulation of electron trajectories in the energy dispersive elements and the lens systems in Chap. 2. We continue with a discussion of some basic properties of the ideal cylindrical field and cylindrical deflectors with equipotential entrance and exit apertures. The optimum aperture angles, the proper match to the lens systems between the monochromator and the analyser, and dispersion compensated spectrometers are considered here. In Chap. 4 we treat the mathematics of the ideal cylindrical field without and also with space charge and show that simple analytically solvable models exist for the trajectories in electron beams in the presence of their own space charge. On the basis of this two-dimensional analysis, the fundamental first-order optical properties of the radial image in a cylindrical deflector are derived as a function of the feed current and other parameters characterising the feed beam and the geometry of the deflector. In parallel with the analytical treatment, we also present two-dimensional computer simulations of the trajectories in the presence of their own space charge. This analytical treatment provides us with an analytical expression for the monochromatic current achievable with a cylindrical deflector and, more importantly, likewise provides us with basic concepts for the design of a spectrometer that is optimised with respect to the space charge limitations. Three-dimensional numerical simulations for realistic deflectors are then described in Chap. 5. These simulations include the "retarding" deflector, which is particularly useful for the first stage in a two-stage monochromator. Important differences between the three-dimensional case and the two-dimensional model will be pointed out.

Chapter 6 is then devoted to suitable emission systems. Having learned in the previous chapters what are the requirements on the dimensions and angular apertures of the feed beam of a monochromator, we are now in a position to specify a cathode emission system. Like the monochromator itself, the beam parameters of cathode emission systems are determined by the space charge. Our treatment of the emission system therefore includes a numerical analysis of trajectories in the presence of space charge. Essential features of the computer codes for the three-dimensional trajectory calculations are explained and the mechanical layout of the cathode emission system is described.

The development of lens systems between the monochromator and the target and between the target and the analyser requires specification of the necessary range of impact energies at the sample and the momentum space, i.e. the acceptance angles there also. In Chap. 7 we therefore briefly consider the various possible applications of electron energy loss spectrometers and the specifications for the beam parameters at the target that arise in the various applications and because of the different scattering mechanisms in electron-surface scattering. A section on the systematics of lens aberrations of nonrotationally symmetric lenses follows. Three different lens systems of high performance are then described. The final chapter, 8, is devoted to a comparison of the theoretical results on space-charge-limited currents in monochromators and the properties of lens systems to actual measurements performed on spectrometers, designed according to the principles and optimisation procedures suggested by the theoretical analysis and the computer simulations.

2. The Computational Procedures

Electron optics involves analytical calculations as well as the numerical simulation of electron trajectories. With high performance computer work-stations readily available, analytical calculations, in particular the use of cumbersome perturbation theories, are being increasingly replaced by numerical studies. This chapter outlines the basic concepts and numerical methods employed in this volume.

2.1 General Strategy

The first approach to a complex electron optical system such as the electron energy loss spectrometer is to identify sections of the device that may be treated as separate objects with respect to the solution of the Laplace equation. To a good approximation, these computationally separate objects are those which are separated by a plate of constant potential, which intersects the beam vertically leaving only a small aperture for the beam. Penetration of the electric field from one space into the other may be neglected in such cases. A typical electron energy loss spectrometer is shown in Fig. 2.1. It consists of a cathode emission system, one (or more) energy dispersive elements for the "monochromator", a lens system between the monochromator and the sample, a second lens system between the sample and the analyser, a further energy dispersive element (the "analyser"), and finally the electron detector. Separate objects of an electron spectrometer are the electron emission system up to the entrance slit at the first monochromator and each of the energy dispersive cylindrical deflectors. We note that our design uses "real" slits for the energy dispersive elements, as opposed to "virtual" slits [2.1]. This is partly because of the computational simplification that follows from the use of real slits. A further advantage is that the resolution of the energy dispersive elements is independent of the potentials applied to the preceding or subsequent lens system and also, more importantly, independent of the image aberrations of the lens systems. On the other hand, real slits in monochromators are subjected to a large current load, which may cause charging of the slit plates. Whether this charging effect is detrimental to the goal of achieving the highest possible monochromatic current remains to be investigated. We will deal with this question in Chaps. 5 and 8.

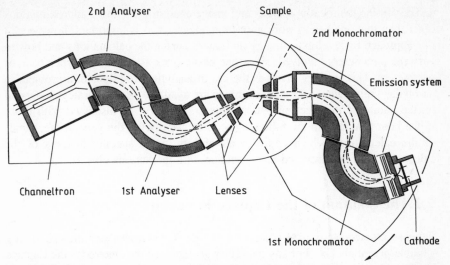

Fig. 2.1. A typical electron spectrometer comprising a cathode emission system, a first energy dispersive system (the monochromator), two lens systems, a second energy dispersive system (the analyser), and an electron detector. Since the energy dispersive elements have small entrance and exit slits, field penetration may be neglected. The cathode system, the energy dispersive systems and the lens systems may then be treated as separate electron optical entities, though their optical properties have to be matched to each other for the optimum performance of the entire system

In addition to the energy dispersive elements terminated by real slits, the lens system between the exit slit of the monochromator and the sample is an electron optical object which can be treated separately, although one has no slit near the sample. The region around the sample, however, is encapsulated within equipotential electrodes, in order to create a field-free region around the sample and ensure well-defined scattering parameters in the experiments. The lens system between sample and the entrance slit of the analyser is likewise a separate object. Frequently, the same type of energy dispersive elements are used for monochromator and analyser. Thus beam parameters as energy, shape and angular apertures are then also similar at the exit slit of the monochromator and at the entrance slit of the analyser. Since, furthermore, the energy loss of electrons scattered from the sample is typically a small fraction of the beam energy at the sample, the same lens system may be used between the sample and the analyser and between the monochromator and the sample. Time reversal invariance of the trajectories ensures that a beam emerging from the exit slit of the monochromator and forming an image of this exit slit at the sample projects an isomorphic image of the exit slit of the monochromator at the entrance slit of the analyser, provided that the lens system is symmetric around the sample in geometry and applied potentials.

Our computational approach for calculating the electron optical properties in each of the separate objects is basically of the "brute force" type, namely, we solve the Laplace equation in two or three dimensions, calculate the trajectories and determine the focal properties and image aberrations with no approximations other than that one has to work with finite elements. It is important, however, in this approach to develop a strategy on how to pursue the calculations and how to vary the parameters involved, since one easily goes astray in the hyperspace of parameters. This strategy involves the use of analytical solutions of a problem in particular limits, whenever such solutions are available. As the strategies depend on the nature of the electron optical objects, they will be discussed with those in the later chapters. In the following sections we discuss the general mechanics of the calculations which apply to all objects, while specific features of the calculation will be discussed in connection with the various objects.

2.2 The Solution of the Laplace Equation

Suppose an electron optical object consists of N electrodes with independently variable potentials eU_i. For any particular geometry of the electrodes the Laplace equation must then be solved N times in order to construct the potential for an arbitrary combination of potentials eU_i. This is achieved by solving the Laplace equation with the potential on all electrodes set equal to zero, save for one electrode i to which one unit of potential, $1\,\mathrm{eV}$ for example, is applied. If the particular solutions of the Laplace equation with these potentials are $V_i(x, y, z)$, then the general solution for an arbitrary choice of all the U_i is

$$V(x, y, z) = \sum_{1}^{N} U_i V_i(x, y, z) \ . \tag{2.1}$$

The linearity of the Laplace equation then guarantees that one has indeed a solution of the Laplace equation since the boundary conditions on the electrodes are correctly fulfilled with the procedure. Furthermore, the uniqueness theorem ensures that the solution is the correct and the only one. The numerical solution of the Laplace equation is performed using a simple algorithm, which is derived from a Taylor expansion of the potential around a particular point in space (x, y, z):

$$V(x + \Delta x, y, z) = V(x, y, z) + \frac{\partial V}{\partial x} \Delta x + \frac{1}{2} \frac{\partial^2 V}{\partial x^2} (\Delta x)^2 \ ,$$

$$V(x - \Delta x, y, z) = V(x, y, z) - \frac{\partial V}{\partial x} \Delta x + \frac{1}{2} \frac{\partial^2 V}{\partial x^2} (\Delta x)^2 \ ,$$

$$V(x, y + \Delta y, z) = V(x, y, z) + \frac{\partial V}{\partial y} \Delta y + \frac{1}{2} \frac{\partial^2 V}{\partial y^2} (\Delta y)^2 \ ,$$

$$V(x, y - \Delta y, z) = V(x, y, z) - \frac{\partial V}{\partial y} \Delta y + \frac{1}{2} \frac{\partial^2 V}{\partial y^2} (\Delta y)^2 \ ,$$

$$V(x, y, z + \Delta z) = V(x, y, z) + \frac{\partial V}{\partial z} \Delta z + \frac{1}{2} \frac{\partial^2 V}{\partial z^2} (\Delta z)^2 \ ,$$

$$V(x, y, z - \Delta z) = V(x, y, z) - \frac{\partial V}{\partial z} \Delta z + \frac{1}{2} \frac{\partial^2 V}{\partial z^2} (\Delta z)^2 \ . \tag{2.2}$$

Adding these equations and using the Laplace equation

$$\frac{\partial^2 V}{\partial x^2} + \frac{\partial^2 V}{\partial y^2} + \frac{\partial^2 V}{\partial z^2} = 0 \tag{2.3}$$

leads to

$$V(x, y, z) = \frac{1}{2\left(\dfrac{1}{(\Delta x)^2} + \dfrac{1}{(\Delta y)^2} + \dfrac{1}{(\Delta z)^2}\right)}$$
$$\times \left(\frac{V(x + \Delta x, y, z) + V(x - \Delta x, y, z)}{(\Delta x)^2}\right.$$
$$+ \frac{V(x, y + \Delta y, z) + V(x, y - \Delta y, z)}{(\Delta y)^2}$$
$$\left. + \frac{V(x, y, z + \Delta z) + V(x, y, z - \Delta z)}{(\Delta z)^2}\right) . \tag{2.4}$$

Repeated application of this equation to each point within a mesh, subject to the appropriate boundary conditions, eventually converges to the numerical solution of the Laplace equation. The final result of several iterations of the algorithm in (2.4) is then an array of numbers describing the potential at each point of the grid when a unit potential is applied to one particular electrode. The final potential grid for starting the trajectory calculation then follows from N such converged arrays using (2.1).

The rapidity of the convergence of the procedure depends on the number of points in the grid and may be speeded up by performing the calculation on a successively finer mesh [2.2]. Whether or not the solution of the Laplace equation has converged sufficiently is best tested by looking at the calculated trajectories, since even for a fixed geometry the number of iterations needed for a sufficiently accurate result depends on the potentials and the nature of the image aberrations one is interested in.

Convergence is also speeded up by using positive feedback [2.3]. Instead of replacing the potential at a given point of the grid by the right hand side of (2.4) one replaces the previous value of the potential $V_{old}(x, y, z)$ by the new value $V_{new}(x, y, z)$ according to

$$V_{new}(x, y, z) = V_{old}(x, y, z) + \text{feedback} \times [V_{calc}(x, y, z) - V_{old}(x, y, z)] . \tag{2.5}$$

A feedback value of 1 reproduces the standard method of calculation. Feedback values above 1 lead to faster convergence. The price to be paid is that the results start to oscillate around their converged values. The procedure becomes unstable when the feedback approaches the value 2. The optimum feedback value has to be established in a set of test runs.

For the cylindrical condensers, the Laplace equation in cylindrical coordinates (r, θ, z) is more appropriate,

$$\frac{\partial^2 V}{\partial r^2} + \frac{1}{r}\frac{\partial V}{\partial r} + \frac{1}{r^2}\frac{\partial^2 V}{\partial \theta^2} + \frac{\partial^2 V}{\partial z^2} = 0 . \tag{2.6}$$

There the algorithm reads

$$V(r, \theta, z) = \frac{1}{2\left(1 + \dfrac{(\Delta r)^2}{r^2(\Delta\theta)^2} + \dfrac{(\Delta r)^2}{(\Delta z)^2}\right)} \left\{ [V(r, \theta + \Delta\theta, z) + V(r, \theta - \Delta\theta, z)] \right.$$

$$\times \frac{(\Delta r)^2}{r^2(\Delta\theta)^2} + V(r + \Delta r, \theta, z)\left(1 + \frac{\Delta r}{2r}\right) + V(r - \Delta r, \theta, z)$$

$$\times \left(1 - \frac{\Delta r}{2r}\right) + [V(r, \theta, z + \Delta z) + V(r, \theta, z - \Delta z)]$$

$$\left. \times \frac{(\Delta r)^2}{(\Delta z)^2} + \varrho(r, \theta, z)\frac{(\Delta r)^2}{\varepsilon_0} \right\} . \tag{2.7}$$

Since we are interested in solving for the potential in the presence of space charge as well, we have added the space charge term $\varrho(r, \theta, z)(\Delta r)^2/\varepsilon_0$, where ε_0 is the vacuum dielectric constant. We note, however, that the simple superposition principle (2.1) does not hold if $\varrho \neq 0$. Also, the feedback is not to be applied to the term containing ϱ.

2.3 Electron Trajectories

Once the Laplace equation has been solved for the unit potentials on each of the independent electrodes, one needs to calculate the trajectories as a function of the potentials on these electrodes. This is performed by a stepwise integration of the Lagrange equations in the appropriate coordinate system. In cartesian coordinates the resulting difference equations are particularly simple. Let $x(t)$, $y(t)$ and $z(t)$ be the coordinates of the electron at the time t; then

$$\dot{x}(t + \Delta t) = \dot{x}(t) + \mathcal{E}_x(x, y, z)\Delta t$$
$$x(t + \Delta t) = x(t) + \dot{x}(t)\Delta t + \tfrac{1}{2}\mathcal{E}_x(x, y, z)\Delta t^2 \tag{2.8}$$

for $x(t)$, with corresponding equations for the y and z coordinates.

$\mathcal{E}_x(x, y, z)$ denotes the electric field at the position x, y, z of the electron and Δt is the time step used in the integration. Note that e/m, the ratio of the electron charge to its mass, may be set equal to unity if one is interested in the shape of the electron trajectories and not in the actual time scale involved.

Some consideration is needed on how to determine the local field at some arbitrary point (x, y, z) from the potential calculated on a grid. The simplest approach would be to take the field to be the difference between the potential on the two nearest grid points along the x-direction and to divide by the distance. This procedure, however, introduces artificial discontinuities in the field, which is not consistent with the fact that the solution of the Laplace equation is smooth up to the second derivative at any point inside the electrodes. The trajectories also become rather inaccurate, unless one uses a very fine grid where the linear dimension of the mesh is small compared to the dimensions of the apertures of the electron optical elements, e.g. the slit-width of the monochromator and the size of its image at the sample. Working with such a small mesh is rather inconvenient, however. Firstly, the convergence of the algorithm used to solve the Laplace equation becomes rather slow. Secondly, one also exceeds memory space of small- and medium-sized computers when three-dimensional calculations are required. It is therefore preferable to use a comparatively coarse mesh with linear dimensions of the order of the typical size of the smallest aperture and calculate the field from the potential with a more elaborate procedure that avoides discontinuities and is still fast enough to permit the calculation of the field at each time interval Δt in a sufficiently short time period. We describe such a method, which we have found to produce fast and accurate results, in the following.

We denote the coordinates of the instantaneous position of the electron where we wish to determine the field for the next step of the integration by (x, y, z). The points on the grid where the potential is known from the numerical solution of the Laplace equation are denoted by coordinates x_i, y_j, z_k, where i, j, k are integers. Suppose that the position of the electron (x, y, z) is such that

$$x_i < x < x_{i+1} , \quad y_j < y < y_{j+1} , \quad z_k < z < z_{k+1} . \tag{2.9}$$

For the next step of the numerical integration we need the components of the electric field vector in all three dimensions. We explain the procedure with the x-component $\mathcal{E}_x(x, y, z)$ as an example. As a first step, we fit the potential along the edges of the bar surrounding the instantaneous position of the electron as depicted in Fig. 2.2. In a second step we calculate the electric field as a function of x on each of the edges of the bar namely $\mathcal{E}_x(x, y_j, z_k)$, $\mathcal{E}_x(x, y_{j+1}, z_k)$, $\mathcal{E}_x(x, y_j, z_{k+1})$, $\mathcal{E}_x(x, y_{j+1}, z_{k+1})$, as the derivative of the polynomial fit to the potential. The field at the actual point of interest (x, y, z) then follows from a linear interpolation.

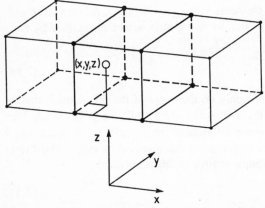

Fig. 2.2. Illustration for the interpolation scheme used to calculate the x-component of the electric field at the instantaneous position $r = (x, y, z)$ of an electron using the potential on 16 points of the grid on which the Laplace equation has been solved numerically

We now describe the mechanics of the procedure in greater detail and begin with the polynomial fit. The order of the polynomials is chosen such that a unique fit to four or six grid points is obtained. Here, we explain the procedure with a third-order polynomial fitted to the four grid points along the edges of the bar as depicted in Fig. 2.2. We use the third-order polynomial

$$V(x, y_j, z_k) \equiv V(x) = a_0 + a_1(x - x_{i-1})$$
$$+ a_2(x - x_{i-1})^2 + a_3(x - x_{i-1})^3. \tag{2.10}$$

Sometimes fitting to a sixth-order polynomial with even exponents only is appropriate, in situations where symmetry requires the coefficients of terms with odd exponents to vanish, as we shall see.

With $\Delta x = x_{i+1} - x_i$ we have a set of equations that determine the coefficients $a_0 - a_3$.

$$V(x_{i-1}) = a_0 ,$$
$$V(x_i) \quad = a_0 + a_1 \Delta x + a_2(\Delta x)^2 + a_3(\Delta x)^3 ,$$
$$V(x_{i+1}) = a_0 + a_1 2\Delta x + a_2 2^2(\Delta x)^2 + a_3 2^3(\Delta x)^3 , \tag{2.11}$$
$$V(x_{i+2}) = a_0 + a_1 3\Delta x + a_2 3^2(\Delta x)^2 + a_3 3^3(\Delta x)^3 .$$

These equations may be cast into matrix form:

$$\begin{pmatrix} V(x_{i-1}) \\ V(x_i) \\ V(x_{i+1}) \\ V(x_{i+2}) \end{pmatrix} = \begin{pmatrix} 0^0 & 0^1 & 0^2 & 0^3 \\ 1^0 & 1^1 & 1^2 & 1^3 \\ 2^0 & 2^1 & 2^2 & 2^3 \\ 3^0 & 3^1 & 3^2 & 3^3 \end{pmatrix} \begin{pmatrix} a_0(\Delta x)^0 \\ a_1(\Delta x)^1 \\ a_2(\Delta x)^2 \\ a_3(\Delta x)^3 \end{pmatrix} \tag{2.12}$$

or in symbolic notation

$$V = M \cdot \tilde{a} , \tag{2.13}$$

where V is a vector having the elements $V(x_{i-1}) \ldots V(x_{i+2})$, $\tilde{a}$ is a vector with the elements $a_0(\Delta x)^0 \ldots a_3(\Delta x)^3$ and M is a matrix whose elements M_{ij} have the form

$$M_{ij} = (i)^j . \tag{2.14}$$

The inverse of this matrix depends only on the order of the polynomial, that is, on the number of points along the x-direction used to fit the polynomial. This number of course remains the same throughout the calculation of a trajectory and the inverse of M therefore needs to be calculated only once. The fitting procedure is thus reduced to a simple matrix multiplication:

$$\tilde{a} = M^{-1} \cdot V . \tag{2.15}$$

The field at x is then calculated from the coefficients of $\tilde{a}$

$$\mathcal{E}_x = a_1 + 2a_2(x - x_{i-1}) + 3a_3(x - x_{i-1})^2 . \tag{2.16}$$

The same procedure is adopted for the three remaining edges of the x-bar. The field at the electron coordinates (x, y, z) is then calculated using the linear interpolation

$$\mathcal{E}_x(x, y, z) = \mathcal{E}_x(x, y_j, z_k) \left(1 - \frac{y - y_j}{\Delta y}\right) \left(1 - \frac{z - z_k}{\Delta z}\right)$$
$$+ \mathcal{E}_x(x, y_{j+1}, z_k) \frac{y - y_j}{\Delta y} \left(1 - \frac{z - z_k}{\Delta z}\right)$$
$$+ \mathcal{E}_x(x, y_j, z_{k+1}) \left(1 - \frac{y - y_j}{\Delta y}\right) \frac{z - z_k}{\Delta z}$$
$$+ \mathcal{E}_x(x, y_{j+1}, z_{k+1}) \frac{y - y_j}{\Delta y} \frac{z - z_k}{\Delta z} . \tag{2.17}$$

The y and z components of the electric field are calculated using the same procedure by fitting the potential along the y and z bars. The number of points and thus the order of the polynomials need not be the same in all three directions. When applied to the calculations of the trajectories in a lens system, one may make use of the fact that the electron trajectories are not too far off the optic axis and that the lens has C_{2v} symmetry, the optic axis being the twofold axis. It is then convenient to take the optical axis as one coordinate axis, typically the x-axis. Because of the C_{2v} symmetry involved, the potential is an even function of the cartesian coordinates y and z, when the y- and z-axes are oriented such that the xy-plane and xz-plane are the mirror planes of the lens sytem. We will come back to this issue and describe some further details of the calculations and appropriate measures to speed up the calculations in Chap. 6.

The trajectory calculations in the cylindrical deflectors may be carried out by a similar procedure. In cylindrical coordinates, the Lagrangian reads

$$L = \frac{m}{2}(\dot{r}^2 + r^2\dot{\theta}^2 + \dot{z}^2) - eV(r, \theta, z) \tag{2.18}$$

where r, θ, z are the radial, the angular, and the z-coordinate orthogonal to the deflection plane, respectively. The stepwise integration of the equations of motion follows the scheme

$$\ddot{\theta}(t) = -2\dot{r}(t)\dot{\theta}(t)/r(t) + \mathcal{E}_\theta(t) ,$$
$$\ddot{r}(t) = r(t)\dot{\theta}^2(t) + \mathcal{E}_r(t) ,$$
$$\ddot{z}(t) = \mathcal{E}_z(t) ,$$
$$\dot{r}(t + \Delta t) = \dot{r}(t) + \ddot{r}(t)\Delta t ,$$
$$r(t + \Delta t) = r(t) + \dot{r}(t)\Delta t + \tfrac{1}{2}\ddot{r}(t)\Delta t^2 ,$$
$$\dot{\theta}(t + \Delta t) = \dot{\theta}(t) + \ddot{\theta}(t)\Delta t ,$$
$$\theta(t + \Delta t) = \theta(t) + \dot{\theta}(t)\Delta t + \tfrac{1}{2}\ddot{\theta}(t)\Delta t^2 ,$$
$$\dot{z}(t + \Delta t) = \dot{z}(t) + \ddot{z}(t)\Delta t ,$$
$$z(t + \Delta t) = z(t) + \dot{z}(t)\Delta t + \tfrac{1}{2}\ddot{z}(t)\Delta t^2 . \tag{2.19}$$

Here $\mathcal{E}_r, \mathcal{E}_\theta, \mathcal{E}_z$ are the derivatives of the potential with respect to r, θ, and z, respetively. As in (2.8), we have set e/m equal to 1. The last two equations of motion need to be integrated only when the dimensions of the deflector in the z-direction are comparable to the size of the gap between the inner and outer deflecting plate so that the terminating potentials at the top and bottom of the deflector become important. For a cylindrical deflector where the gap is small compared to the height, the $\mathcal{E}_z$-component of the field vanishes and the trajectories become straight lines with respect to z. As we shall see in the next chapters, the use of a vertical height of the deflector comparable to a gap size with an adjustable potential on the top and bottom cover plates offers the possibility of adjusting the first-order focus via this potential and also of balancing the divergence of the beam due to electron-electron repulsion.

The interpolation of the field components $\mathcal{E}_r, \mathcal{E}_\theta, \mathcal{E}_z$ is performed as described above with a modification concerning $\mathcal{E}_z$: fitting the potential around the centre plane of the deflector is performed with an even polynomial in the z-coordinate measured from the central radial plane. It is thus assumed that top and bottom plate have the same potential, which makes the deflector symmetric with respect to the central plane.

Finally, it should be noted that the interpolation scheme for the field requires the potential to be defined one mesh unit before the starting point of the electron e.g. at the entrance slit of the deflector (2.10) and also one mesh unit beyond the target. In our calculations, which neglect field penetration through the slits, we take these potentials to be equal to the potentials at the starting and target position, respectively.

2.4 Space Charge Limited Current

Since the prime objective of this study was to improve the intensity of high resolution electron energy loss spectrometers, space charge will be an important consideration. While it is obvious that space charge places an upper limit on the current of a monochromatic beam, it is less obvious which of the electron optical elements involved in the production of the beam is the limiting factor. In this section we will give some qualitative consideration to this issue.

Electron trajectories in the presence of space charge and likewise the maximum current which may be injected into an aperture cannot be calculated in a closed form except for a few simple geometries. Since at this point we are merely interested in the order of magnitude, it should suffice to estimate the current using one of the simple geometries for which the space charge problem can be solved analytically, and use this solution as a crude representation of the actual electron optical system.

We begin the analysis with the estimate of the maximum current which can be fed into the entrance slit of the monochromator. For the space charge problem, we use the model of the space charge limited current flow between two parallel plates [2.4]. In an actual cathode emission system, the cathode has the form of a fine tip and furthermore there are focusing elements between the cathode and the entrance slit of the monochromator. Such a device allows the beam to diverge between the cathode and the entrace slit and the effect of the space charge in a real cathode emission system is therefore less than for the parallel plate arrangement. The maximum feed current calculated for the parallel plate system is therefore a lower bound for the current of a normal cathode emission system. For a parallel plate system the current that may pass through the entrance slit of the monochromator is

$$I_{\text{in}} = \frac{4}{9} k E^{3/2} \frac{hs}{d_k^2} \tag{2.20}$$

where h is the height of the entrance slit, s its width, d_k the distance of the emitting plate from the entrance aperture, and E the energy of the electrons at the entrance aperture of the monochromator. For reasons of simplicity it is assumed that the initial kinetic energy of the emitted electrons is zero. The space charge constant k is

$$k = \varepsilon_0 \sqrt{\frac{2}{me^2}} = 5.25 \times 10^{-6} \text{ A/eV}^{3/2} . \tag{2.21}$$

As already mentioned, the value given by (2.20) is a lower bound to the maximum feed current. An alternative to the parallel plate model is to consider the broadening of a ribbon-shaped beam under the influence of its own space charge. This model has been discussed elsewhere [Ref. 2.5, p.36]. It eventually leads to the same equation for the maximum feed current as (2.20), except that the numerical factor is now two, rather than 4/9. It thus appears that (2.20) is a reasonable, model-independent, analytical description of the monochromatic current.

Inspection of (2.20) tells us that the effect of space charge in the cathode system does not appear to impose any upper bound on the maximum feed, since with respect to the cathode emission system one is free to choose appropriate values for the parameters h, s, d_k and E in (2.20). As, on the other hand, one finds experimentally that the current produced by a monochromator is indeed limited, the monochromator itself appears to be the current-limiting electron optical object. Technical improvements, which would permit the monochromator to handle more feed current, should therefore directly generate higher count rates in a spectrum. In order to achieve such an improvement a detailed numerical analysis of the optical properties of the cylindrical condenser in the presence of space charge was performed and will be presented in Chap. 5. In the end, the nature of the cathode emission system will also determine the performance of the spectrometer, since the monochromator and transport lenses also require that the feed beam for the monochromator be well collimated.

The significance of the space charge in the energy-dispersive parts of an electron spectrometer has been noted earlier [2.1], and an analytical expression for the monochromatic current of a cylindrical deflector was deduced from a simple model, which accounts for the effect of space charge on the electron trajectories inside the cylindrical deflector [Ref. 2.5, pp. 49ff.]. While this model was the first to predict monochromatic currents which were roughly consistent with experimental results, it failed to provide correct guidelines for the optimisation of cylindrical deflectors to operate under space charge conditions, as we shall see.

3. The Electron Optics of the Cylindrical Deflector

The basic properties of the cylindrical deflector as an electron dispersive device are discussed in this chapter. It is shown that cylindrical deflectors terminated by equipotential plates have no disadvantages compared to an ideal field termination when the deflection angle is properly adjusted. Cylindrical deflectors can also be used as retarding devices, which allows a stepwise monochromatisation of a beam, without intermediate lenses for retardation. The transmission of electron dispersive systems is related to the angular aberration. The concept of "dispersion compensated" spectrometers and the reasons for their failure in practice are investigated.

3.1 The Ideal Cylindrical Field

In this section we summarise a few basic equations for the ideal cylindrical field and electron trajectories in that field. The equations permit the calculation of the electron optical properties and the energy resolution of devices with an ideal cylindrical field. In practical systems, the field is distorted due to the presence of entrance and exit apertures, which are usually equipotential surfaces. The fringe field of the cylindrical deflector is therefore quite different from the ideal cylindrical field. We will see in Sect. 3.2 that the influence of the fringe field on the electron optical properties is by no means a marginal one. The equations for the electron trajectories in the ideal field are therefore of limited practical use. They do serve, however, as a reference frame for the assessment of the electron optical properties of a real deflector. The presentation to follow also introduces basic parameters and their notation.

The electric field between two metal cylinders of infinite length is given by

$$\mathcal{E}_r = \frac{\Delta V}{\ln(R_2/R_1)} \frac{1}{r} , \tag{3.1}$$

where $\mathcal{E}_r$ is the radial component of the electric field and ΔV the voltage difference between the cylinders. R_2 and R_1 are the inner radius of the outer cylinder and the outer radius of the inner cylinder, respectively; r is the radial coordinate.

An electron with mass m and velocity v in the deflecting field $\mathcal{E}_r$ travels on a circle of radius r when

$$\frac{mv^2}{r} = \frac{2E_0}{r} = e\mathcal{E}_r , \tag{3.2}$$

where E_0 is the "pass energy" of the electron and e the elementary charge. The trajectories of electrons subjected to the field of an ideal cylindrical deflector may be calculated analytically [3.1] (see also Sect. 4.1). It is straightforward to show, e.g., that an electron optical device with the ideal cylindrical field has first-order focusing properties with respect to the entrance angle α_1, where α_1 denotes the angle in the radial plane between the tangent to the circle of radius r and the actual electron trajectory. If θ denotes the angular coordinate of an electron in the cylindrical field, first-order focusing is achieved after the electron has travelled a distance equivalent to the angular coordinate

$$\theta_{\text{f, ideal}} = \frac{\pi}{\sqrt{2}} \approx 127.28° \ . \tag{3.3}$$

If electrons embark on trajectories at a particular angular coordinate $\theta = 0$ with an angle α_1 and with a small radial deviation y_1 from a particular radius r_0 (typically the central path), then y_2, the radial deviation at the first-order focus point at $\theta = \theta_{\text{f, ideal}}$, will be given by

$$y_2 = -y_1 + r_0 \left(\frac{\delta E}{E_0} \right) - \frac{4}{3} r_0 \alpha_1^2 \ , \tag{3.4}$$

where δE is the deviation of the electron energy from the pass energy E_0 (3.2). Thus, the ideal cylindrical field shows an energy dispersion equivalent to the mean radius r_0

$$E_0 \frac{dy}{dE} = r_0 \ . \tag{3.5}$$

Equation (3.4) also tells us that the magnification of a cylindrical field optical device is -1, which means that when a slit is placed so as to limit the maximum value of y_1, the image of this slit will be of the same width. The position of the image depends on the energy, which makes the device an energy selective one when equipped with an entrance and an exit slit. From (3.4) and (3.5) we may also calculate the energy resolution of the device. According to (3.4), the maximum positive energy deviation δE_+ from the pass energy E_0 is permitted for an electron with $\alpha_1 = 0$ which passes the entrance slit at $y_1 = +s/2$ and arrives at the exit slit with $y_2 = +s/2$, where s is the width of the entrance and exist slit,

$$\frac{\delta E_+}{E_0} = \frac{s}{r_0} \ . \tag{3.6}$$

When the entrance slit is illuminated with a beam consisting of a bundle of electron trajectories with a distribution of angles α_1 up to a maximum angle α_{1m} the device transmits electrons of nonzero α_1 at even higher δE_+. The maximum positive energy deviation δE_+ is then

$$\frac{\delta E_+}{E_0} = \frac{s}{r_0} + \frac{4}{3} \alpha_{1m}^2 \ . \tag{3.7}$$

16

The maximum negative energy deviation δE_- for a transmitted electron occurs when the electron passes the entrance slit with an angle $\alpha_1 = 0$ at $y_1 = -s/2$ and arrives at the exit slit with $y_2 = -s/2$. Inserting this condition into (3.4) yields

$$\frac{\delta E_-}{E_0} = -\frac{s}{r_0} \; . \tag{3.8}$$

The base width of the transmitted electron beam is therefore

$$\frac{\Delta E_B}{E_0} = \frac{2s}{r_0} + \frac{4}{3}\alpha_{1m}^2 \; . \tag{3.9}$$

This is the basic equation describing the energy resolution of an ideal cylindrical deflector. Similar equations hold for other deflecting energy dispersive devices [3.2]. The effect of the second-order angular aberration term in (3.4) on the energy resolution is quite obvious. While the base width of the transmitted energy distribution is easily calculated, the full width at half maximum $\Delta E_{1/2}$ is usually the quantity which is quoted for an energy selective device. If α_{1m} is small, the transmitted energy distribution has a triangular shape with the full width at half maximum of

$$\Delta E_{1/2} = \tfrac{1}{2}\Delta E_B \; . \tag{3.10}$$

This result simply follows from the -1 magnification of the device which makes the transmission function a self-convolution of the rectangularly shaped transmission function of a slit aperture. For larger α_{1m}, the transmitted energy distribution develops a round shape and also acquires a tail towards the high energy side. We shall deal with this issue in greater detail in Sect. 3.4. It is useful to define a measure of the angular spread of the injected beam in terms of the ratio s/r_0. With the help of (3.4), one finds that the maximum entrance angle α_1 for an electron having the nominal pass energy ($\delta E = 0$) is

$$\alpha_{1m} = \left(\frac{3s}{4r_0}\right)^{1/2} \; . \tag{3.11}$$

Electrons injected into a cylindrical deflector with $\alpha > \alpha_{1m}$ merely contribute to the (high-energy) tail of the transmitted energy distribution and thus have an adverse effect on the performance.

In the derivation of (3.4), an analytical form for the electron trajectories in the ideal cylinder field was used in which radial deviations y from the mean radius r_0 and angular deviations α_1 up to second order were considered. An analytical calculation of high order terms, though feasible, is a major task, and numerical analysis of electron trajectories becomes advantageous. This is even more true when the field differs from the ideal cylindrical field, which is the typical practical situation. We have performed the numerical solution of the Laplace equation in two dimensions using the algorithm (2.7). Comparison with the analytical expression for the potential serves to test the program and the convergence of the result. The potential was calculated using a grid of 200 points along the

17

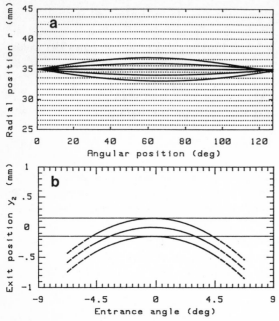

Fig. 3.1. (a) Electron trajectories with the initial conditions $\alpha_1 = -4.5°$, $-2.25°$, 0, $+2.25°$ and $4.5°$ and $r = r_0 = 35\,\mathrm{mm}$ in the ideal cylindrical field. The dotted lines are the equipotential lines. The angular coordinate is displayed from 0 to 127.2°, the radial coordinate from $R_1 = 25\,\mathrm{mm}$ to $R_2 = 45\,\mathrm{mm}$. (b) The radial deviation y_2 at the exit slit ($\theta = 127.2°$) as a function of the entrance angle α_1. The parameter is the radial deviation at the entrance slit $y_1 = -0.15$, 0, $+15\,\mathrm{mm}$. The figure displays y_2 between -1 and $+1$. The parallel lines indicate a reasonable size for the slits, namely $s = 0.3\,\mathrm{mm}$. The curves for y_2 are displaced parabolas indicating that no significant higher order corrections to (3.4) occur for the ideal cylindrical field

angular coordinate θ and 100 points on the radial coordinate r, with the radius of the inner and outer electrode $R_1 = 25\,\mathrm{mm}$ and $R_2 = 45\,\mathrm{mm}$, respectively. The trajectories were calculated using 5000 integration steps. The trajectories – as calculated for the ideal cylinder field – are shown in Fig. 3.1a, where the radial coordinate r and the angular coordinate θ are displayed as cartesian coordinates. The trajectories focus after a pass length corresponding to a deflection angle of $\theta_f = 127.28°$. In Fig. 3.1b, the radial deviation y_2 at $\theta = 127.28°$ is plotted as a function of the entrance angle α_1. The radial deviations y_2 appear to be parabolas displaced by y_1 just as described by (3.4). This is confirmed when a polynomial is fitted to the calculated values of $y_2(y_1, \alpha_1)$. One may use a polynomial fit of the type

$$y_2(y_1, \alpha_1) = C_y y_1 + C_{yy} y_1^2 + C_{yyy} y_1^3 \ldots + C_\alpha \alpha_1 + C_{\alpha\alpha} \alpha_1^2 + C_{\alpha\alpha\alpha} \alpha_1^3 + \ldots$$
$$+ C_{\alpha y} \alpha_1 y_1 + C_{\alpha yy} \alpha_1 y_1^2 + C_{\alpha\alpha y} \alpha_1^2 y_1 \ldots . \qquad (3.12)$$

For the range of interest here, $|y_1| < 1\,\mathrm{mm}$, $|\alpha_1| < 3°$, we find that all coefficients but C_y and $C_{\alpha\alpha}$ are negligible ($10^{-2} - 10^{-3}$ of the C_y, $C_{\alpha\alpha}$ terms). For C_y and $C_{\alpha\alpha}$ we obtain

18

$$C_y = -1.00 , \tag{3.13}$$

$$C_{\alpha\alpha} = -46.66 \,\text{mm} \approx -\tfrac{4}{3}r_0 , \tag{3.14}$$

as predicted by the analytical solution. The coefficient C_{yy} and higher orders appear to vanish identically.

It may be useful at this stage to compare the aberrations of the ideal cylindrical field with the aberrations of rotationally symmetric lenses and relate our terminology to that used there. Lens aberrations are classified according to the expansion (3.12). The term linear in α vanishes because the expansion is performed at the first-order focal point where the coefficient C_α vanishes by definition. The coefficient C_y is the magnification. It is straightforward to show that all the second-order expansion coefficients vanish also, if y denotes the deviation from the optic axis and if the lens has circular symmetry, or at least a mirror plane perpendicular to the direction y. The lowest aberration coefficients are of third order in α_1 and y_1. The effect of the four third-order aberration coefficients on the image is illustrated in Fig. 3.2a–d. We begin with the third-order term in α_1 as shown in Fig. 3.2a. Beams with larger α_1 have a shorter focal length than beams with smaller α_1. Because of the symmetry of the lens, this shortening effect must be even in $\alpha(\sim \alpha^2)$ which makes the lowest coefficient in Δy_2 of third order in α_1 (Fig. 3.2a). For electrostatic lenses, the third-order coefficient is always negative [3.3]. For a rotationally symmetric lens, one therefore defines a spherical aberration by

$$\Delta y_2 = -C_s\alpha_1^3 \quad \text{with} \quad C_s > 0 . \tag{3.15}$$

The "coma" aberration is illustrated in Fig. 3.2b: beams of larger α emerging from points off the optical axis form a stigmatic image with a smaller magnification than beams with small α.

Off-axial points have their first-order image on a non-planar surface. Figure 3.2b shows the image surface for rays within the "meridional" plane spanned by the optical axis and the y-direction. Rays within the "sagittal" plane perpendicular to the meridional plane may have their focus on a differently curved image surface. If the two image surfaces have a different curvature then a stigmatic focusing cannot be achieved. This lens aberration is therefore named astigmatism.

Finally, one has the distortion as the last of the third-order aberrations (Fig. 3.2d). The effect of distortion is that the image of an object in the form of a square become distorted to a pin-cushion- or a barrel-shaped image, depending on the sign of the distortion coefficient. If we compare the classification of the image aberrations developed for circular lenses and in ordinary light optics with the aberrations that we have encountered in the optical properties of the ideal cylindrical field, we find that the image formation in the ideal cylindrical field is free of coma, astigmatism and distortion. It is also free of third-order angular aberration. This third-order aberration is, however, replaced by an angular aberration of second order in α, an aberration that vanishes identically for a rotationally symmetric lens.

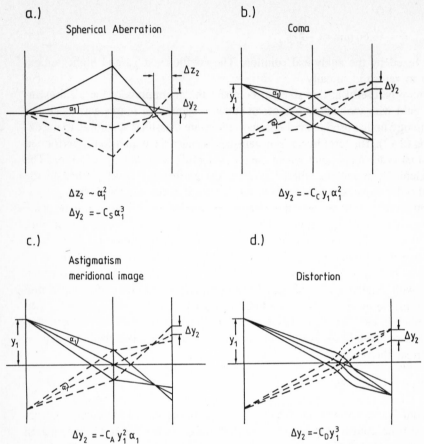

a.)

Spherical Aberration

$$\Delta z_2 \sim \alpha_1^2$$
$$\Delta y_2 = -C_S \alpha_1^3$$

b.)

Coma

$$\Delta y_2 = -C_C y_1 \alpha_1^2$$

c.)

Astigmatism
meridional image

$$\Delta y_2 = -C_A y_1^2 \alpha_1$$

d.)

Distortion

$$\Delta y_2 = -C_D y_1^3$$

Fig. 3.2a–d. Illustration of the four different third-order aberrations in the image. Rays drawn as dashed lines are derived from the rays drawn as solid lines through a mirror operation. (a) Angular aberration, also known as "spherical" aberration for rotationally symmetric lenses. The mirror symmetry (or the rotational symmetry) requires that the lowest angular aberration coefficient is of third order in α_1. (b) The coma aberration. The coma may be described as the dependence of the magnification on α_1. (c) Astigmatism in the meridional plane. For off-axial points, the first-order image in the angle α_1 lies on a curve which is a circle in lowest order. Extended to three dimensions the curve becomes a surface known as the meridional image surface. On a flat image plane the α_1-image of an off-axial point becomes a line. The first-order focal point for rays inclined with respect to the meridional plane lies on a different image surface, the tangential image surface. In between the tangential and the meridional image surface, one may define a surface of least confusion for the image of an off-axial point object. (d) The distortion may be understood as the dependence of the magnification on the size of the object. The distorted image of a square is either pin-cushion or barrel shaped, depending on the sign of the distortion coefficient

We notice from this comparison of aberrations in the ideal cylindrical field and in lenses that the classification and terminology developed for lenses are not necessarily useful in the assessment of energy dispersive devices because of the different symmetry of the optical elements. There is a further deeper rooted difference. In light optics, a Taylor expansion of the image aberrations and a classification of aberrations according to this expansion serves a useful purpose as higher order expansion coefficients are small. In electron optics, especially in the electron optics of electrostatic elements, the higher order aberration coefficients are by no means small. It would therefore be pointless to try to optimise a cylindrical deflector by modifications that affected the second-order angular aberration without monitoring the higher order expansion coefficients. More appropriate methods specific to the electron optical elements and objectives of their design will be discussed in the sections to follow.

3.2 The Cylindrical Deflector Terminated with Equipotential Electrodes

In order to use the cylindrical deflector as an energy selective device, entrance and exit apertures are required. When these apertures take the form of a metal plate with a slit, the potential inside the cylindrical deflector is substantially distorted near the entrance- and exit-plate. The distortions of the fringe fields may be minimised by replacing the solid metal plate by a series of small metal strips bearing each a potential chosen to match the logarithmic potential inside the deflector. Such a fringe field correction is typically applied in cylindrical mirror type analysers as used for Auger- and other electron spectroscopies. Because of the complications introduced by such field terminators it seems worthwhile to search for alternatives and also to investigate the electron optical properties of cylindrical deflectors with equipotential terminating apertures.

A relatively simple correction to the fringe field is achieved by dividing the terminating electrode into three parts, with the potential of the central part matched to the central potential of the cylindrical deflector and those of the inner and outer electrodes to the potential of the inner and outer deflecting electrodes, respectively. Figures 3.3a and b show equipotential lines for a cylindrical deflector terminated by a single radial metal plate and by a divided plate as described above. One sees that for electrons near the central path, the field remains nearly the same as that of the ideal cylindrical field, save for a small fraction of the path length where the electron approaches the aperture closely. The technical realisation of a potential termination as in Fig. 3.3b is obviously rather simple as no additional potentials need to be applied. Nevertheless, we have decided not to use this type of potential termination because it involves large field gradients in the vicinity of the electron beam, which cause large deflections of trajectories of electrons having an energy different from the pass energy. In combination with surface charging of the electrodes caused by the electron beam, one may create

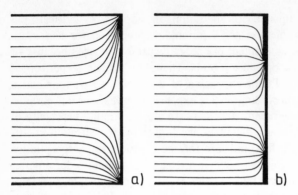

Fig. 3.3. (a) Equipotential lines for a cylindrical deflector terminated by a metal aperture. The potential of the aperture is the arithmetic average between the potential of the inner and outer deflecting plate. The figure refers to a ratio $R_2/R_1 = 1.8$. As with previous figures the radial and angular coordinates are displayed as cartesian coordinates. (b) A simple fringe field correction is obtained when the aperture is divided and the potentials of the inner and outer deflecting electrodes are applied to the inner and outer parts of the aperture, respectively. The field penetrates less into the deflector than in Fig. 3.3a

a situation that may be difficult to control. In the following we will therefore investigate the optical properties of the cylindrical deflector with terminating metal plate electrodes.

The Laplace equation was again solved according to the algorithm (2.7). An analytical expression for the potential in a cylindrical deflector terminated by equipotential plates [3.4] is known but this has the form of a Fourier-series, which converges very slowly. For the calculation of the trajectories, the derivatives of the potential with respect to the cylindrical coordinates are needed. There, the Fourier-series does not converge at all. This circumstance was overlooked in a previous treatment [3.5, 6] of the trajectories in a terminated deflector.

The Laplace equation was solved on a 100×200 grid, where the numbers refer to the radial and angular coordinates, respectively. The convergence of the Laplace algorithm is slower than for the ideal cylindrical field. As before, we tested whether the result for the potential had converged sufficiently by calculating the trajectories. The results were found to be satisfactory after 2000 iterations of the Laplace algorithm over the entire field, when no feedback was used. In these calculations the potential on the terminating electrodes was equal to the average of the potential on the inner and outer deflecting plate.

Figure 3.4 shows the trajectories for a cylindrical deflector with the inner and outer radius of $R_1 = 25\,\text{mm}$ and $R_2 = 45\,\text{mm}$, respectively. The radial coordinate at the entrance position was $r_0 = 35\,\text{mm}$. A first-order focus is achieved but at a significantly smaller deflecting angle than with the ideal cylindrical field. Instead of $\theta_f = 127.3°$ we now obtain $\theta_f \approx 106.8°$. Figure 3.4b again shows the exit position y_2 as a function of the entrance angle α_1, with the starting position $y_1 = -0.15, 0, +0.15\,\text{mm}$ as a parameter. Comparison of Fig. 3.1b and 3.4b shows that the image aberrations for the terminated deflector are qualitatively similar to

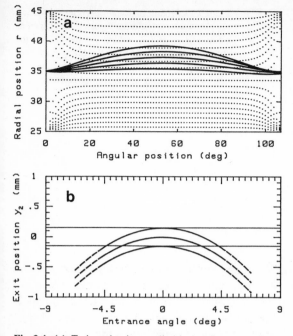

Fig. 3.4. (a) Trajectories in a cylindrical deflector terminated by equipotential entrance and exit apertures. First-order focus is achieved at an angle $\theta_f = 106.8°$ when the radii of the inner and outer plates are $R_1 = 25\,\text{mm}$ and $R_2 = 45\,\text{mm}$, respectively. The initial radial coordinate r_0 for the trajectories is in the centre between the inner and outer deflection plates, $r_0 = (R_2 + R_1)/2$. The potential on the apertures is the arithmetic mean between the potential of the inner and outer deflecting plate. Note that in the cylindrical field this mean potential is found at the radial coordinate $r_m = \sqrt{R_2 R_2}$, which is smaller than r_0. The effect of the aperture potential and the initial radial coordinate on the first-order focal length is relatively minor, though not negligibly small. (b) Radial position y_2 at the first-order focus as a function of the entrance angle α_1. The parameter is the initial radial position $y_1 = -0.15, 0, 0.15\,\text{mm}$. The parallel lines indicate a slit of 0.3 mm. The y_2-curves are parabolas displaced by 0.15 mm, indicating that the magnification is -1 and that the largest aberration term is the second-order angular aberration

those of the ideal cylindrical field, the α_1^2-term being again the main aberration. A quantitative analysis of the aberration terms shows that the α_1^2-term is about 10% larger than for the ideal field. Specifically we find

$$C_{\alpha\alpha,\,\text{real}} = -1.48\, r_0 = 1.11\, C_{\alpha\alpha,\,\text{ideal}} \ . \tag{3.16}$$

The quantiative analysis indicates also that the coefficients C_{yy}, $C_{y\alpha}$ and $C_{y\alpha\alpha}$ are nonzero. For all practical purposes these aberrations are too small to have a significant effect on the transmission and the transmitted energy distribution. It is important to note that the magnification of the device remains -1 and that the exit angles α_2 centre around zero when the entrance angles do. Finally, the energy dispersion is somewhat less than for the ideal field. Here we find

$$E_0 \frac{dy}{dE} = 33.8\,\text{mm} = 0.966\, r_0 \ . \tag{3.17}$$

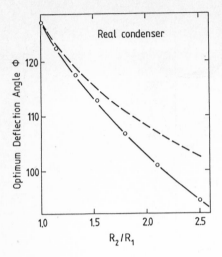

Fig. 3.5. Deflection angle subtended by the deflecting plates where first-order focusing is obtained as a function of R_2/R_1, where R_2 and R_1 are the radius of the outer and inner deflecting plates, respectively. The dashed line is the optimum deflection angle according to the Herzog correction

The slightly larger angular aberration and the smaller energy dispersion seem to indicate a deterioration of the electron optical properties of the cylindrical deflector when terminated by metal plate apertures. However, for a fair comparison one needs to consider also the smaller deflection angle, i.e. the reduced size of the device. If one normalises a deflector with the ideal cylindrical field and the real deflector as described above, so that they have the same angular aperture and the same path length, the resolution of the terminated deflector is actually slightly better than the ideal field deflector. Thus the rather convenient scheme of terminating the cylindrical deflector by metal entrance and exit apertures has no adverse effect on the electron optical properties.

The reduction of the first-order focal length for the terminated cylindrical deflector depends on the size of the gap between the inner and outer deflecting plate, or more precisely on the ratio of the inner and outer radius. In Fig. 3.5, we plot the deflection angle for first-order focusing θ_f as a function of the ratio R_2/R_1. The angle θ_f approaches 127.3° as R_2/R_1 approaches unity. The reason is that, as the gap between the inner and outer cylinder closes, the fringe field is more and more confined to the immediate vicinity of the aperture plate. Thus the fraction of the electron path length in which the fringe field affects the electrons becomes smaller. In calculating the trajectories for the terminated deflector, we have assumed that the aperture plate is oriented along the radial coordinate, and that the gap between the aperture plate and the inner and outer deflecting plate is equivalent to an angle of 127.3/200 degrees. Cylindrical deflectors may also be terminated by an aperture plate placed at some distance from the deflecting plates. The appropriate corrections to the deflecting angle were first calculated by *Herzog* [3.7] who used the fringe field solution for the parallel plate condenser to estimate the fringe field correction of the cylindrical deflector. More extensive analytical calculations have also been performed by *Wollnik* and *Ewald* [3.2], but there again only an approximate solution for the fringe field was used which, like the Herzog correction, is applicable only in the limit of small gaps between the

24

inner and outer deflecting plates. This is also demonstrated in Fig. 3.5, where the reduction of the deflection angle according to the Herzog correction is plotted as a dashed line. While the Herzog correction leads to a practically identical result for small gaps, i.e. when R_2/R_1 is nearly one, there is a significant discrepancy for larger gaps.

For high resolution electron energy loss spectroscopy, one is interested in larger gaps for several reasons. Larger gaps require larger deflecting voltages to be applied to the deflecting plates. The optical properties of the deflector should therefore become less subject to spurious potentials caused by local variations of the work function. Secondly, a larger gap also avoids the problem of the reflection of electrons of false energy from the deflecting plates. Such electrons, by virtue of this deflection, may pass through the exit slit, despite having an energy quite different from the pass energy. This causes spurious loss- or gain-peaks in the spectra, a feature which plagued earlier instruments [3.8]. For gap sizes corresponding to a ratio of $R_2/R_1 = 1.8$, as used here in our model deflector, the spectra are largely free of such spurious peaks. So far we have investigated the effect of a terminating plate, the potential of which is the arithmetic average of the potential of inner and outer deflecting plates. The deflector, however, also works quite well with a different potential on the aperture plate. One merely has a moderate shift in the first-order focal length as a function of this potential. The focal length becomes larger when the potential on the aperture is varied more in the negative sense. This may be used for a fine tuning of the focal length in order to compensate for imperfections in the geometry or deviations caused by charging and spurious work function effects. More importantly, the deflector may also be used as an asymmetric, retarding device. An example is shown in Fig. 3.6a. The nominal pass energy of the device is 1 eV, and 1 eV is also the potential of the electrons at the entrance aperture. The exit aperture however is at 1/3 eV, which means that the electrons leave the device with the energy reduced by a factor of three. The first-order focus is now at $\theta_f \sim 111°$ when $R_2/R_1 = 1.8$, as before (compare Fig. 3.4). The exit angle of the trajectories may again be centred around zero, if one allows for a slight offset in the radial position of the entrance and exit slits, r_{01} and r_{02}, the entrance slit being shifted inwards (by 1.2 mm when $r_0 = 35$ mm) and the exit slit shifted outwards by the same amount. Unlike the symmetric deflector, the asymmetric deflector has a magnification different from one. Empirically we found the approximate relation

$$C_y \approx -F^{0.25} , \tag{3.18}$$

where F is the retardation factor (3 in the example shown here). In order to achieve optimum transmission, the width of the exit slit s_2 should match the width of the image of the entrance slit, which is $|C_y|s_1$. The angular aperture of the exit beam is also different from the aperture of the entrance beam. This is in fact required from the conservation of phase space. Applied to a bundle of trajectories which have an entrance aperture angle α_1 and an exit aperture angle α_2, the conservation of phase space in two dimensions requires

25

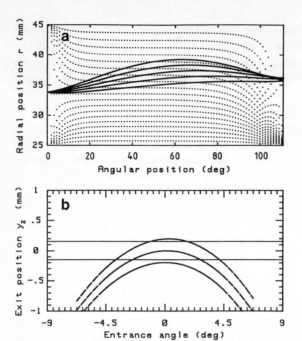

Fig. 3.6. (a) Trajectories in a retarding deflector. The entrance aperture is at the average potential as before (Fig. 3.4), the potential on the exit aperture is reduced by a factor of three. A first-order focus is again achieved, now at $\theta_f = 111°$. The radial coordinate at the entrance (r_{01}) and the exit aperture (r_{02}) are displaced from the centre by about $r_{01} - r_0 = -1.2\,\text{mm}$ and $r_{02} - r_0 = 1.2\,\text{mm}$, when $r_0 = 35\,\text{mm}$. This is in order to achieve a mean exit angle α_2 of about zero when the entrance angles α_1 are also centred around zero. (b) The radial coordinate $y_2 = r - r_{02}$ at the exit slit as a function of the entrance angle α_1. Parameter y_1 as in Fig. 3.4b. The second-order angular aberration is larger than for the symmetric deflector, and so is the energy dispersion. The magnification is also larger than one (~ 1.3)

$$\alpha_1 y_1 \sqrt{E_1} = \alpha_2 y_2 \sqrt{E_2} \ . \tag{3.19}$$

With $E_2 = E_1/3$ and using (3.18), one obtains

$$\alpha_2 = \alpha_1 F^{0.25} \ . \tag{3.20}$$

As seen from Fig. 3.6b, the retarding deflector again has a second-order angular aberration. For the model deflector shown here with $F = 3$ we obtained

$$C_{\alpha\alpha} = 2.05\, r_0 \ . \tag{3.21}$$

The angular aberration is larger than for the symmetric deflector (3.16). This larger aberration must be matched against the energy dispersion which is also larger. For the deflector with $F = 3$ we found

$$E_0 \frac{dy}{dE} = 48\,\text{mm} = 1.37\, r_0 \ . \tag{3.22}$$

The ratio of the angular aberration to the energy dispersion is thus comparable

26

for the symmetric and the retarding deflector. As we shall see later, retarding deflectors are extremely useful devices for forming monochromatic beams of high intensity. The final monochromator is usually preceded by a deflector operating at higher pass energy. It has been realised for some time that the use of such pre-monochromators leads to higher monochromatic currents [3.9, 10]. However, previous designs such as that described in [3.10] use a relatively complex retarding lens between the pre-monochromator and the monochromator. Our calculation indicates that such a lens is superfluous when pre-monochromator and monochromator are properly designed and matched to each other. This remark is also pertinent to monochromators in the presence of space charge, as we shall see.

We conclude this section with an extension of the analysis into the third dimension, retaining however the assumption of a strictly 2D cylindrical field. It is not particularly difficult to obtain an essentially 2D potential in the centre of a cylindrical deflector. It is merely necessary to ensure that the deflecting plates and the aperture plates extend sufficiently far into the third dimension, along the z-axis. Typically a total height of the deflector of four times the size of the gap between the inner and outer deflecting plate is sufficient. Focusing then occurs only in the radial plane, while along the z-direction the trajectories are straight lines. This is true, provided that the heights of the entrance and exit slits are sufficiently small to keep all beams within the centre part of the cylindrical deflector, where the influence of the top and bottom fringe fields is negligible. An upper bound on the height of the slits is also advisable for another reason. Energy selection in the cylindrical field is governed by the radial components of the velocity. An electron traversing the cylindrical field on a trajectory inclined at an angle β to the radial plane has an additional z-component of the velocity

$$v_z = v_\theta \tan \beta , \tag{3.23}$$

where v_θ is the velocity along the central path.

The kinetic energy of this electron is hence

$$E = \frac{m}{2} \left(v_z^2 + v_\theta^2 \right) \approx \frac{m}{2} v_\theta^2 (1 + \beta^2) ,$$
$$E = E_0(1 + \beta^2) , \tag{3.24}$$

which is larger than the pass energy E_0 by an amount

$$\Delta E \approx E_0 \beta^2 . \tag{3.25}$$

This term gives rise to additional broadening of the transmitted energy distribution. The use of a particular slit height limits the maximum value of β and thus the unwanted broadening of the energy distribution.

3.3 Transmission of the Cylindrical Deflector

In the preceding sections we have seen that the image formation of the ideal cylindrical field as well as the properties of the cylindrical deflector terminated by equipotential apertures may be described by

$$y_2 = +C_y y_1 + 2D \left(\frac{\delta E}{E_0} - \beta^2 \right) + C_{\alpha\alpha} \alpha_1^2 , \qquad (3.26)$$

$$z_2 = z_1 + r_0 \theta_f \beta_1 , \qquad (3.27)$$

where D is the energy dispersion and z_1 and z_2 are the vertical positions in the object and image plane, respectively. The remainder of the notation is as before. Using (3.26) and (3.17), one may calculate analytically the transmission function of the deflector as a function of the incident energy, the angular aperture of the feed beam α_1, β_1 and the geometric parameters of the device such as the width and the height of the entrance and the exit aperture, denoted by s_1, s_2, h_1, h_2, respectively. The calculation does however involve rather elaborate, though elementary, integrations [3.11] and the result does not readily furnish a simple recipe for constructing an optimised device. Numerical analysis is another possibility, and we will perform such an analysis in the next section. As numerical calculation of a transmission function involves the integration of several thousands of trajectories, a substantial computational effort would be required to lucidate the dependence of the transmission on the many parameters of the system. It is therefore useful to perform a simplified analysis of transmission including consideration of the base width of the transmitted energy distribution and then discuss the various procedures that yield an optimum set of geometrical parameters for monochromator and analyser.

We first consider the transmission at the nominal pass energy where $\delta E = 0$. Figure 3.7a illustrates the exit slit of the deflector together with the size and position of the image of the entrance slit. In Fig. 3.7a and in the following it is assumed that the width of the image of the entrance slit $|C_y| s_1$ is smaller than or equal to the width of the exit slit s_2. We allow the magnification $|C_y|$ to be different from unity, which covers the normal as well as the retarding deflector. Because of the second-order angular aberration term in (3.27), the image is shifted by $C_{\alpha\alpha} \alpha_1^2$. As long as the entrance angle α_1 is small enough for the image of the entrance slit to fit into the exit slit, the transmission is unity. For entrance angles larger than a critical angle

$$\alpha_{t_1} = \left(\frac{s_2 - |C_y| s_1}{2 |C_{\alpha\alpha}|} \right)^{1/2} \qquad (3.28)$$

the transmission decreases according to the fraction of the image that fits into the exit slit. Thus for the transmission as a function of α_1, we have

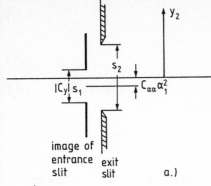

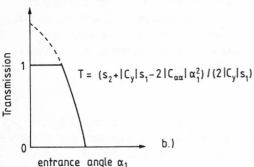

Fig. 3.7. (a) Illustration of the size and position of the (first-order) image of the entrance slit of a cylindrical deflector at the exit slit. The image is smaller than the entrance slit if $|C_y|s_1 < s_2$, which is the case shown here. The image is shifted by the amount $C_{\alpha\alpha}\alpha_1^2$. (b) Transmission (i.e. the fraction of the electrons that pass through the exit slit) as a function of the entrance angle α_1. Electrons are assumed to have the nominal pass energy

$$T(\alpha_1) = \begin{cases} 1 , & 0 \le \alpha_1 \le \alpha_{t_1} , \\ \dfrac{s_2 + |C_y|s_1 - 2|C_{\alpha\alpha}|\alpha_1^2}{2|C_y|s_1} , & \alpha_{t_1} \le \alpha_1 \le \alpha_{t_2} , \end{cases} \tag{3.29}$$

where

$$\alpha_{t_2} = \left(\frac{s_2 + |C_y|s_1}{2|C_{\alpha\alpha}|} \right)^{1/2}$$

is the critical angle beyond which all electrons with the nominal pass energy are blocked by the exit aperture. The transmission function $T(\alpha_1)$ is displayed in Fig. 3.7b when $|C_y| < 1$. As one is essentially interested in electrons with an energy near the pass energy, we see already that the angular aperture of the feed beam α_{1m} should not exceed α_{t_2}. For the simple yet important case where $|C_y| = 1$ and $s_1 = s_2$, the total transmission T_α of a deflector, when fed with a beam having a homogeneous angular distribution between $\pm\alpha_{1m}$, may be calculated easily. The result is

$$T_\alpha = \begin{cases} 1 - (|C_{\alpha\alpha}|/3s)\alpha_{1m}^2 , & \alpha_{1m}^2 < s/|C_{\alpha\alpha}| , \\ \frac{2}{3}\alpha_{1m}^{-1}\left(s/|C_{\alpha\alpha}|\right)^{1/2} , & \alpha_{1m}^2 > s/|C_{\alpha\alpha}| . \end{cases} \tag{3.30}$$

The base width of the transmitted energy distribution is also easily calculated using (3.26). Following the same procedure as in Sect. 3.1 one has

$$\frac{\delta E_-}{E_0} = -\frac{(s_2 + |C_y|s_1 + 2|C_{\alpha\alpha}|\alpha_{1m}^2)}{2D}, \tag{3.31}$$

$$\frac{\delta E_+}{E_0} = +\frac{(s_2 + |C_y|s_1)}{2D} + \beta_{1m}^2 \tag{3.32}$$

for the minimum and the maximum energy distribution, respectively. The total base width of the transmitted energy distribution is then

$$\frac{\Delta E_B}{E_0} = \frac{s_2 + |C_y|s_1 + |C_{\alpha\alpha}|\alpha_{1m}^2}{D} + \beta_{1m}^2. \tag{3.33}$$

With these equations one may establish a simple criterion for a cylindrical deflector, optimised with respect to the angular apertures α_{1m} and β_{1m}. For simplicity we now confine ourselves to a symmetric deflector for which $s_2 = s_1$ and $C_y = -1$. As we shall see in the next chapter, the current at the detector of a spectrometer approximately scales according to

$$I_D \sim \Delta E_{1/2\,tot}^n, \tag{3.34}$$

where $\Delta E_{1/2\,tot}$ is the FWHM of the energy distribution transmitted by the spectrometer (usually referred to as the "resolution"). The exponent n is a number of the order of 2–3. Since the total current that may be fed into a deflector is proportional to the entrance apertures α_{1m}, β_{1m}, the quantity

$$P = \frac{\alpha_{1m}\beta_{1m}}{\Delta E_B^n} \tag{3.35}$$

may be adopted as a figure of merit for the performance of the deflector. In the performance factor we disregard the loss in transmission described by (3.30), since the resulting optimum angle is comparatively small. After inserting (3.33) for ΔE_B, it is easy to calculate the maximum value of the performance relative to α_{1m} and β_{1m}; the optimum angular apertures are found to be

$$\alpha_{1m\,opt}^2 = \frac{s}{|C_{\alpha\alpha}|}\frac{1}{n-1}, \tag{3.36}$$

$$\beta_{1m\,opt} = \frac{s}{D}\frac{1}{n-1}. \tag{3.37}$$

For the ideal cylindrical deflector these equations reduce to

$$\alpha_{1m\,opt} = \left(\frac{3s}{4r_0(n-1)}\right)^{1/2}, \tag{3.38}$$

$$\beta_{1m\,opt} = \left(\frac{s}{r_0(n-1)}\right)^{1/2}. \tag{3.39}$$

One may use these equations as a first guide for choosing the appropriate relations between s, r_0, α_{1m}, and β_{1m} and also the slit height. Modifications to these simple considerations will be brought about by the lens system and the space charge in the monochromator. These equations also show that the optimum angle α_{1m} is smaller than the greatest angle that would still allow electrons to be transmitted at the pass energy.

In a spectrometer, the monochromator also has to match the lenses and the analyser. The monochromatic current is projected onto the sample via a lens system which images the exit slit of the monochromator onto the sample. As we have discussed before, conservation of phase space imposes certain relations between the energy, the angular aperture, and the size in the object and image plane. We assume that the xy- and the xz-planes are mirror planes of the lens system, the x-axis being the optic axis. Conservation of phase space may then be formulated separately in the two mirror planes

$$\alpha_{2m} s_2 \sqrt{E_0} = \alpha_{is} s_{is} \sqrt{E_{is}} , \tag{3.40}$$

$$\beta_{2m} h_2 \sqrt{E_0} = \beta_{is} h_{is} \sqrt{E_{is}} . \tag{3.41}$$

Here α_{2m} and β_{2m} are the maximum angles of the beam emerging from the monochromator in the xy-plane and xz-plane, respectively. They are equal to the maximum angles of the feed beam of the monochromator α_{1m} and β_{1m}, provided that the deflector has symmetric potentials on the entrance and exit apertures, where one has also $C_y = -1$. The suffix "is" refers to the corresponding quantities of the incident beam on the sample. In deriving (3.40) and (3.41) it is also assumed that α and β be small, so that $\sin \alpha \sim \alpha$. By virtue of the electron beam excitation, the sample emits electrons whose characteristic energy is typically near the impact energy. We assume these electrons to have an even distribution in momentum space. The current of these electrons is then proportional to

$$I_e = \begin{cases} \sigma \alpha_{es} \beta_{es} s_{es} h_{es} I_M , & s_{es}, h_{es} \leq s_{is} h_{is} , \\ \sigma \alpha_{es} \beta_{es} I_M , & s_{es} h_{es} \geq s_{is} h_{is} , \end{cases} \tag{3.42}$$

where the quantities with the suffix es now refer to the characteristic quantities of the trajectories emerging from the sample and σ is the excitation probability pertinent to the type of excitation to be studied. I_M is the monochromatic current produced by the monochromator, which we will find to scale to the base width of the monochromator (Sect. 4.4) according to

$$I_M \sim \Delta E_{BM}^{5/2} . \tag{3.43}$$

It follows immediately from (3.42) that the area at the sample that is imaged into the entrance aperture of the analyser should match the dimensions of the illuminated area on the sample. If we now apply the rules of phase space conservation to the process of image formation between sample and analyser, we find for the current I_e

$$I_e = \sigma \frac{\alpha_{1mA}\beta_{1mA}}{E_{es}M_y M_z} E_{0A} I_M , \qquad (3.44)$$

where $\alpha_{1mA}, \beta_{1mA}$ are the maximum angles of trajectories into the analyser in the xy- and xz-planes, respectively and M_y, M_z are the linear magnifications of the lens system in the y and z directions, respectively, and E_{0A} is the pass energy of the analyser. It is convenient to have an expression for the current at constant resolution, and we therefore replace E_{0A} and I_M by (3.33) and (3.43) and obtain the following expression for the current in the electron detector after passing the analyser

$$I_D = \sigma T_M T_{L1} T_{L2} T_A \frac{\alpha_{1mA}\beta_{1mA}}{E_{es}M_y M_z}$$
$$\times \frac{1}{\left(2s_A + |C_{\alpha\alpha}|\alpha_{1mA}^2\right)/D + \beta_{1mA}^2} \Delta E_{BA} \Delta E_{BM}^{5/2} . \qquad (3.45)$$

Here ΔE_{BA} and ΔE_{BM} are the base widths of the energy distribution of the analyser and monochromator, respectively. We have also added the product of the four transmission functions, which characterise the transmission of the monochromator (T_M), the first lens system between monochromator and the sample (T_{L1}), the second lens system between sample and analyser (T_{L2}), and the analyser (T_A). Each of these transmission functions, to lowest order in the apertures angles α and β, has the form

$$T \sim 1 - t_\alpha \alpha^2 - t_\beta \beta^2 \qquad (3.46)$$

with a priori unknown coefficients t_α and t_β. It is thus obvious that (3.45) cannot be used to calculate the optimum parameters for the cylindrical deflector without knowing the properties of the lenses involved and nor can these lenses be optimised without reference to properties of monochromator and analyser. Equation (3.45) also tells us that one should try to keep the image small, i.e. use small values of the linear magnifications M_y and M_z, in order to have a large acceptance angle. On the other hand we shall find later that the lens aberrations increase, when the magnification is too small. We must recognise that the optimisation of an electron spectrometer involves an element of skill, despite the advance in computational techniques. It also involves an iterative process in the optimisation of the various elements.

Nevertheless, (3.45) shows us how to match the resolution of the analyser to that of the monochromator. Omitting those factors from (3.45) that do not change to first order when the resolution of the analyser is changed, the current at the detector is proportional to

$$j_D \sim \Delta E_A \Delta E_M^{5/2} T_A . \qquad (3.47)$$

The analyser transmission T_A splits into two factors. One, the term depending on the entrance aperture angle α_{1m}, has already been calculated (3.30). The second arises from the fact that the analyser weights the incoming energy distribution with its own transmission curve with respect to energy. Energy transmission curves are well approximated by gaussians. The transmission factor arising from energy may hence be described by

$$T_{EA} = (2\pi\Delta E_A)^{1/2}(8\ln 2)^{1/4} \int \exp[-4E^2(\Delta E_A^{-2} + \Delta E_M^{-2})\ln 2]dE$$

$$= \left(1 + \left(\frac{\Delta E_M}{\Delta E_A}\right)^2\right)^{-1/2}, \tag{3.48}$$

where ΔE_M and ΔE_A are the FWHM of the analyser and monochromator, respectively. Thus the variation of the current at the detector with the resolution of the analyser is

$$j_D \propto \frac{\Delta E_A^2(\Delta E_{tot}^2 - \Delta E_A^2)^{5/4}}{\Delta E_{tot}}. \tag{3.49}$$

In deriving this result, we have used $\Delta E_{tot}^2 = \Delta E_M^2 + \Delta E_A^2$, which follows from the width of the convolution of two gaussians with the FWHM of ΔE_M and ΔE_A, respectively. The optimum current is obtained when

$$\Delta E_A = \frac{2}{3}\Delta E_{tot} = \frac{2}{\sqrt{5}}\Delta E_M \tag{3.50}$$

and the current is then

$$j \propto \Delta E_{tot}^{7/2}. \tag{3.51}$$

The power 7/2 instead of 5/2 for the monochromatic current arises from the reduction in the acceptance angle at the sample as ΔE_A decreases. If one probes the monochromatic current produced by the monochromator with the analyser directly, in order to test the performance of the spectrometer (Chap. 8), then one power in ΔE_A disappears from the expression for the monochromatic current of the detector and one has

$$j_D \sim \frac{\Delta E_A(\Delta E_{tot}^2 - \Delta E_A^2)^{5/4}}{\Delta E_{tot}} \propto \Delta E_{tot}^{5/2} \tag{3.52}$$

with the optimum match between analyser and monochromator at

$$\Delta E_A = \sqrt{\frac{2}{7}}\Delta E_{tot} = \sqrt{\frac{2}{5}}\Delta E_M. \tag{3.53}$$

The current j_D is plotted as a function of the ratio $\Delta E_A/\Delta E_{tot}$ in Fig. 3.8 for the two cases discussed above.

How could a higher resolution in the analyser be achieved? Keeping in mind that the energy loss at the sample is small, so that it may be neglected to first

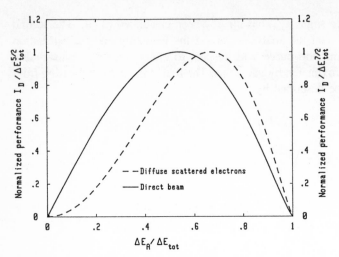

Fig. 3.8. This figure illustrates the current at the detector I_D of an electron spectrometer divided by $\Delta E_{1/2}^{5/2}$ and $\Delta E_{1/2}^{7/2}$, respectively, as a function of the ratio of the resolution of the analyser to the combined resolution of monochromator and analyser $\Delta E_{1/2\,tot}$. The exponent 5/2 corresponds to the case where the monochromatic current is measured directly. The exponent 7/2 refers to electrons emerging from a sample in a diffuse angular distribution. The detector current follows a different power law in the two cases since the acceptance angle at the sample is proportional to $\Delta E_{1/2\,A}$

order, the exit slit of the monochromator is imaged onto the entrance slit of the analyser and the angular and size parameters of monochromator and analyser are thus again connected via phase space conservation

$$s_M \alpha_M \sqrt{E_{0M}} = s_A \alpha_A \sqrt{E_{0A}} , \tag{3.54}$$

$$h_M \beta_M \sqrt{E_{0M}} = h_A \beta_A \sqrt{E_{0A}} , \tag{3.55}$$

where the indices M and A refer to monochromator and analyser, respectively. Let us suppose that monochromator and analyser operate at the same energy and have the same dimensions, angular apertures, and resolution, so that the projected image of the exit slit of the monochromator also matches the size and shape of the entrance slit of the analyser perfectly and the lens system would in theory have a 100% transmission. If one now attempts to lower ΔE_A of the analyser in order to comply with (3.49), the entrance angle α_A or the size of the image s_A or both increase so that the product is increased by a factor of $(5/2)^{1/4}$, which has an adverse effect on either the transmission or the resolution. A remedy is to use an analyser with a larger radius and also a larger slit height.

3.4 Numerical Simulation of the Transmission

In performing the numerical analysis we let ourselves be guided by the considerations that led to the equation for the optimum aperture angles α_{1m} and β_{1m} for the feed beam. The free parameters of the system are then the radius r_0 and the slit width s, or rather the ratio of the two. We have chosen $s = 0.3\,\text{mm}$ and $r_0 = 35\,\text{mm}$.

In the interest of having smaller devices, it may be an advantage to reduce the total height of the deflectors and terminate the deflectors by a top and bottom plate and apply a potential to these plates equivalent to the arithmetic average of the potentials on the deflecting plates. Again, as with the entrance and exit aperture plates, one could correct for the fringe field by appropriate measures. It is however advisable to abstain from such measures as the potential on the top and bottom plate is a valuable adjustable parameter of the system.

In order to simulate the effect of the top and bottom plates, we have performed 3D-potential and trajectory calculations on a $50 \times 100 \times 30$ grid. The numbers refer to the r-, θ- and z-coordinates, respectively. With regard to the z-coordinate one needs to calculate and have available only the upper (or the lower) half of the deflector when symmetric potentials are applied to the top and bottom plates. The interpolation scheme for the field was as described in Sect. 2.3.

One effect of applying a potential to the top and bottom plates is to shift the first-order focus. An example is shown in Fig. 3.9 for a deflector where the total height is about $0.63\,(R_1 + R_2)$. The angle of first-order focusing increases when a negative potential is applied. The shift of the focal length can become quite substantial when large potentials are applied to the top and bottom plates. The reason for this shift becomes apparent when one considers the shape of

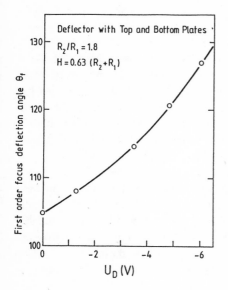

Fig. 3.9. Optimum deflection angle of the cylindrical deflector as a function of the potential on the top and bottom shields U_D. The zero of the potential coincides with the average of the potential on inner and outer deflecting plates. The data were calculated for $R_2/R_1 = 1.8$ and a total height of the device of $H = 0.63(R_2 + R_1)$

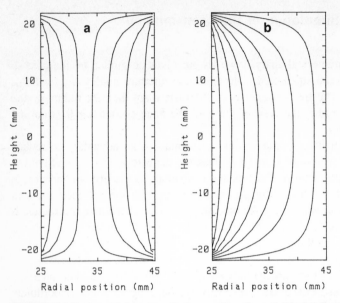

Fig. 3.10. (a) Equipotential contours in the zr-plane of a cylindrical deflector when top and bottom cover plates are at a potential equal to the average of the outer and inner deflection plate. (b) Equipotential contours when the potential of the top and bottom cover plates is equal to the potential on the outer deflection plate. The equipotential contours now resemble those of a spherical deflector

the equipotential lines in the zr-plane for a negative bias on the deflecting plates (Fig. 3.10). The equipotential lines become more and more curved near the center of the deflector as the negative bias on the top and bottom plates is increased. The potential eventually becomes similar to the potential of a spherical deflector for which the deflection angle for first-order focusing is 180°. It is indeed possible to design a pseudo-spherical deflector [3.12] by placing top and bottom shields on a cylindrical analyser. The extra degrees of freedom one has with the potential, the height, and also the shape of the top and bottom shield may be used to optimise the system with respect to particular aberration coefficients. We found however that in general the gain from such an exercise was a relatively minor one. A negative bias potential on the top and bottom plates obviously also has an effect on the shape of the trajectories with respect to the z-coordinate. First-order focusing with respect to the angle β may even occur when the bias is large enough. The pseudo-spherical deflector is indeed characterised by having the same first-order focal length with respect to both the angles α and β, so that one has a stigmatic image of the entrance aperture at the first-order focus. The two-dimensional focusing of the spherical deflector and the pseudo-spherical deflector is usually considered as an advantange over the cylindrical deflector. The issue is however more complex than it may appear at first glance. A detailed assessment of the virtues and drawbacks of the various types of electrostatic analysers requires an understanding of the behaviour of these devices under space charge conditions as well as a knowledge of the fundamental properties of the lens systems between the monochromator and sample and between the sample and the analyser.

In order to illustrate how the divergence of the beam along the z-axis affects the transmission, we show the result of a simulation for a particular analyser. The dimensions of the analyser are as before. The pass energy E_0 was assumed to be $0.5\,\mathrm{eV}$. The analyser was also equipped with top and bottom shielding plates, each placed at a distance of $22\,\mathrm{mm}$ from the centre plane. The potential on the top and bottom plates was set to $-0.3\,\mathrm{V}$ relative to the average between the potential of the inner and outer deflection plate. The entrance slit, $0.3 \times 6\,\mathrm{mm}$ in size, was fed by a beam with a homogeneous angular distribution between the limits $\alpha_m = \pm 3°$ and $\beta_m = \pm 4°$ to satisfy (3.36) and (3.37) approximately. Figure 3.11b displays the effect of the negative bias on the top and bottom plates on the trajectories. When exit and entrance slits are of the same height, the negative bias enhances the transmission of the device, though slightly, while the effect on the focal length and thus on the resolution is marginal for the quoted potential on the top and bottom shields (Fig. 3.11a, c).

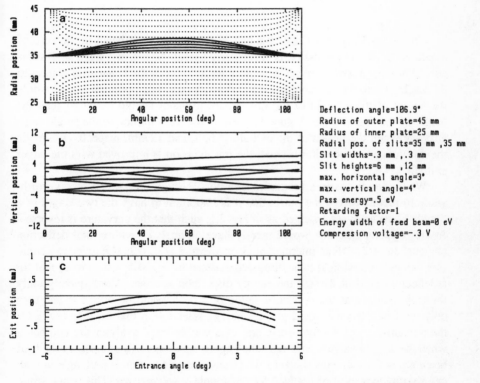

Deflection angle=106.9°
Radius of outer plate=45 mm
Radius of inner plate=25 mm
Radial pos. of slits=35 mm ,35 mm
Slit widths=.3 mm ,.3 mm
Slit heights=6 mm ,12 mm
max. horizontal angle=3°
max. vertical angle=4°
Pass energy=.5 eV
Retarding factor=1
Energy width of feed beam=0 eV
Compression voltage=-.3 V

Fig. 3.11a–c. Trajectories in a cylindrical deflector with equipotential plates as entrance and exit apertures and a negative bias of $-0.3\,\mathrm{V}$ applied to a top and a bottom plates. (a) Trajectories in the radial plane. (b) Trajectories in the tangential plane. (c) Angular aberration with respect to the angle α_1

37

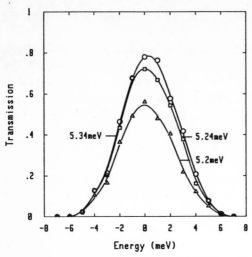

Fig. 3.12. The transmitted energy distribution of the cylindrical deflector in Fig. 3.11. The height of the entrance slit is 6 mm. The circles, squares and triangles refer to different heights of the exit slits of 12, 9 and 6 mm, respectively

The main effect of the negative bias is to compress the beam in the vertical direction. Such a compression is particularly useful when a two-stage analyser is used. The compression voltage then may serve to focus the beam with respect to the angle β into the entrance aperture of a detector. For optimum transmission, the height of the exit slit of the first monochromator should be larger than the height of the entrance slit. Figure 3.12 shows the transmission of the single stage analyser when the height of the exit slit is 6, 9, and 12 mm, respectively. While the difference in resolution is small the transmission is improved with the larger exit slit.

We finally discuss the properties of a double-stage analyser. It seems worthwhile to mention from the outset that it is important to have the two stages of a double-stage analyser arranged as in Fig. 2.1, such that the curvature is reversed. In the opposite case one would merely have an analyser with a total deflection angle of $2\pi/\sqrt{2}$ with an intermediate stop aperture at $\pi/\sqrt{2}$. The trajectory equation in a cylindrical field to be discussed in detail in Sect. 4.1 will show us that for a deflecting field of $2\pi/\sqrt{2}$ the energy dispersion vanishes! Consequently, only the stop aperture at the deflection angle of $\pi/\sqrt{2}$ would be effective and thus only the first analyser would provide energy selection in that case. In Fig. 3.13, the transmission of a single-stage and of a double-stage analyser are compared when the second stage is inverted. The angular apertures α_{1m} and β_{1m} of the feed beam are both assumed to be zero. The transmission function is a triangle in both cases with the resolution doubled for the double-stage analyser. This result, while in accordance with the trajectory equations, is noteworthy insofar as one might have thought that for two sequentially arranged analysers the transmission would be the product of the transmission function of each analyser, as it is for two

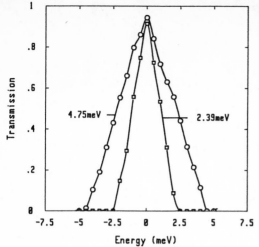

Fig. 3.13. Comparison of the transmitted energy distribution of a single-stage analyser (o) and a double-stage analyser (□) when the angular apertures of the feed beam α_{1m} and β_1 are both zero

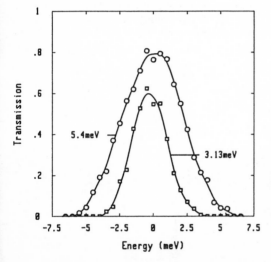

Fig. 3.14. Comparison of the transmitted energy distribution for a single-stage analyser such as in Fig. 3.11 (o) and a double-stage analyser of the same type (□). The angular apertures α_{1m} and β_1 are 3° and 4°, respectively

sequentially arranged optical filters. This is not the case! The reason is that the exit position and energy of the electron are related, so that at the entrance slit of the second analyser the energy distribution as a function of the radial position is not homogeneous as (assumed) for the first analyser.

In the case where the angular apertures are finite ($\alpha_{1m} = 3°$ and $\beta_{1m} = 4°$), the transmission function resembles a Gaussian in both cases considered above (Fig. 3.14). For the double-stage analyser the FWHM is reduced to 0.58 rather than to 0.5. Regardless of the size of α_{1m} and β_{1m}, a double-stage analyser has the same resolution as an analyser of twice the radius and the same slit width when the angular apertures of the feed beam remain the same. This is illustrated in Fig. 3.15, where a double-stage analyser with radii of 35 mm is compared with

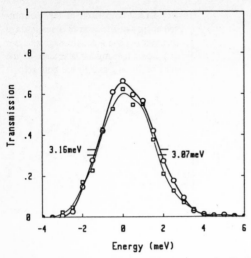

Fig. 3.15. Comparison of the transmitted energy distribution for a double-stage analyser as in Fig. 3.13 (o) and a single-stage analyser of twice the radius (□). The angular apertures α_{1m} and β_1 are 3° and 4°, respectively, for both analyser systems. Although both systems are equal according to this calculation, the background intensity due to secondary electrons may be less for the double-stage system

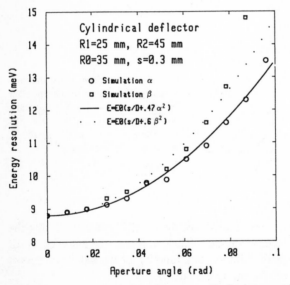

Fig. 3.16. Numerical simulation of the energy distribution transmitted by a cylindrical deflector with equipotential apertures when the angular distributions of the feed beam are of a rectangular shape. The FWHM is plotted versus the aperture angles α_{1m} and β_{1m}

a single-stage analyser with $r_0 = 70 \, \text{mm}$; in both cases, the feed beam has an angular aperture $\alpha_{1m} = 3°$ and $\beta_{1m} = 4°$ as before. The only advantage of the double-stage analyser is therefore that the background intensity due to secondary electrons or electrons scattered from the deflection plates is less. The same effect is however achieved when a second aperture is placed after the exit in order to block electrons leaving the analyser at large angles α, which would be the case for electrons scattered from the deflection plates.

While the base width of the energy distribution can be calculated directly from the dispersion and the angular aberration (3.33), the full width at half maximum (FWHM) can only be obtained from a numerical simulation. We have performed such a simulation assuming rectangular profiles of the angular distribution in the angles α_1 and β_1. The results for the FWHM for 1 eV pass energy are shown in Fig. 3.16. The results are well described by the equation

$$\Delta E_{1/2} = E_0(s/D + 0.47\alpha_{1m}^2 + 0.6\beta_{1m}^2) \,, \tag{3.56}$$

where D is the dispersion, which is $0.966 \, r_0$ for the deflector with equipotential apertures (3.17). On comparing (3.56) and (3.33), one notices that the FWHM is not equal to half the base width.

3.5 Dispersion Compensation Spectrometers

Dispersion compensation spectrometers were first designed and built by *Kevan* and *Dubois* [3.13]. Such spectrometers promise a large enhancement of the throughput by essentially parallel processing of the electrons of the entire energy distribution emitted from the cathode, whereas conventional spectrometers work with only a small fraction of those electrons. The essence of this beautiful idea lies in the exploitation of the relation between energy and position at the exit of a monochromator described above. The relation must be maintained in feeding the analyser, and monochromator and analyser must have the same handedness of the curvature. The relation between energy and exit position y_2 of the monochromator is provided by (3.4)

$$y_2 = -y_1 + r_{01}\frac{\Delta E}{E_0} - \frac{4}{3}r_{01}\alpha_1^2 \,, \tag{3.57}$$

where ΔE, the energy deviation from the pass energy E_0, can assume any value permitted by the width of the energy distribution emitted by the cathode. The quantities y_1, r_{01}, α_1 have their usual meaning. We assume for the moment that there is no lens system between monochromator and analyser, or that the lens system is perfect in the sense that each point y_2 at the exit of the monochromator has an exact stigmatic image at the same radial position at the entrance of the analyser. The entrance angle α of the analyser is then also equal to the exit angle of the monochromator and thus equal to α_1. The analyser transports the beam further according to

$$y_3 = -y_2 + r_{02} \frac{(\Delta E + \delta E)}{E_0} - \frac{4}{3} r_{02} \alpha_1^2 . \tag{3.58}$$

In (3.58) we have allowed for the pass energy of the analyser to be different by a small amount δE. Combining (3.57) and (3.58) one finds for $r_{01} = r_{02} = r_0$

$$y_3 = y_1 + r_0 \frac{\delta E}{E_0} . \tag{3.59}$$

All electrons emitted from the cathode thus appear at the same exit position regardless of their initial energy and also independently of their angle α_1 with respect to the central path. The maximum and minimum energy deviations δE for which electrons still pass through the exit slit of the analyser are

$$\frac{\delta E_+}{E_0} = \frac{s_1 + s_3}{2r_0} \quad \text{and} \tag{3.60}$$

$$\frac{\delta E_-}{E_0} = \frac{s_1 + s_3}{2r_0} , \tag{3.61}$$

respectively, where s_1 and s_3 are the widths of the entrance slit of the monochromator and the exit slit of the analyser. The base width of the transmitted energy distribution is therefore

$$\frac{\Delta E_B}{E_0} = \frac{2s}{r_0} \tag{3.62}$$

with $s = s_1 = s_3$, which is the same as for a single monochromator with no angular aberrations. The widths of the exit slit of the monochromator and the entrance slit of the analyser do not enter at all and can therefore be made broad enough to accommodate the entire energy spectrum of the cathode.

In order to calculate the transmission function vs the energy shift δE, that is, the shape of a spectral line, we again resort to computer simulation. For simplicity, we use equations (3.57) and (3.58), which correspond to the ideal cylindrical field. Fringe field corrections and the nonlinearity of the energy dispersion for larger deviations from the central pass energy are unimportant, though any distortion of the image not compensated by the analyser directly affects the resolution. We shall return to this issue shortly. In Fig. 3.17 the transmission is shown as a function of δE, for various widths s_2 of the exit (entrance) slits of the monochromator (analyser). One sees that the shape of the transmission function remains triangular and independent of the size of the entrance slit s_2. The current at the detector rises with the size of s_2, as long as the monochromator is fed with an energy distribution broad enough to fill the exit slit of the monochromator. The ratio s_2/s_1 is the gain factor in throughput compared to a conventional spectrometer. The shape of the transmission function is also independent of α_1. Unlike the transmission function of a single cylindrical deflector, the shape of a spectral line in a dispersion-compensated spectrometer and the overall resolution are not affected by the quadratic angular aberration term.

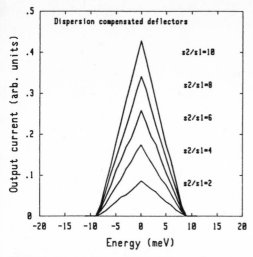

Fig. 3.17. Transmission vs difference in pass energy between monochromator and analyser for an ideal dispersion-compensated spectrometer. The parameter is the width s_2 of the exit slit of the monochromator, which is equal in size to the entrance slit of the analyser. The shape of the transmission curve is independent of s_2 while the throughput increases linearly with s_2, when the monochromator is fed with a broad energy distribution. The width of the entrance slit of the monochromator s_1 and the exit slit of the analyser s_3 are taken to be 0.3 mm, the central radii r_0 as 35 mm. The shape of the transmission function is not affected by the α^2-term

These properties would make the dispersion-compensated spectrometer superior to any other design. A prerequisite for this type of spectrometer to be operational is however that the perfect correlation between energy and position in the exit of the monochromator is maintained in the entrance of the analyser. Normally a lens is required between monochromator and sample, and between sample and analyser in order to allow the impact energy on the sample to be varied. Any distortion of this image of the exit slit of the monochromator projected onto the analyser entrance degrades the resolution. The problem is obviously particularly severe when one attempts to use large gain factors and thus a distortion-free image of a large aperture is needed. The effect of image distortion is simulated in Fig. 3.18 by assuming that a point at the exit slit of the monochromator is spread over a disk at the entrance slit of the analyser. The size of the disk is assumed to be 5% and 10% of the width of the slit s_2, respectively. The gain factor was taken as $s_2/s_1 = 10$. If $s_1 = 0.3$ mm the size of the disk of confusion is thus 0.3 mm for a 10% image distortion. According to our experience with electrostatic lenses this is probably far on the low side, at least when a large acceleration between monochromator and sample and in turn a large retardation towards the analyser is required. Nevertheless, Fig. 3.18 makes it clear that the effect of the quite moderate distortion on the resolution is severe. The base width of the spectral line is nearly doubled for 10% distortion. As the current of a spectrometer is proportional to the 3rd–4th power of the resolution (3.34), the gain factor of ten in the throughput is already lost with a 10% image distortion.

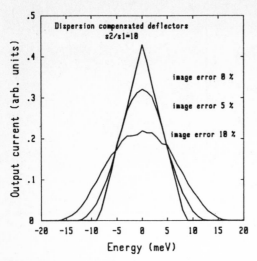

Fig. 3.18. Transmitted energy distribution of a dispersion-compensated spectrometer when the required perfect correlation between the position at the exit slit of the monochromator and the entrance slit of the analyser is distorted, e.g. by the lens system. If the image of a point at the exit of the monochromator is blurred to a disk of 10% of the width of the entrance slit of the analyser, the base width is approximately doubled when the gain factor s_2/s_1 is 10

This is the reason why the disperson-compensation scheme has not lived up to its initial promise. Furthermore, it is easy to see that a dispersion-compensated spectrometer is bound to have a rather variable resolution since the quality of the image depends not only on the fundamental design parameters of the lenses but also on the homogeneity of the surface potentials of the lens elements. Charging and variable work functions of deposits on lens elements place that homogeneity beyond control. It is therefore advisable to adopt spectrometer designs for which the resolution to first order does not change with the inhomogeneity of surface potentials. Finally, no version of the principle of dispersion compensation in the presence of space charge, that is, for high currents, is known.

4. The Electron Optics
of the Ideal Cylindrical Field with Space Charge

The monochromatic current is limited by the increasing electron-electron repulsive forces in high density beams. A simple analytical model for the effect of the "space charge" on the electron optical properties of cylindrical deflectors is presented. First-order compensation of the space charge effect is obtained by enlarging the deflection angle and by applying an additional negative bias to the top and bottom shields terminating the cylindrical deflectors. In spherical deflectors the azimuthal and radial foci are displaced from each other for high current loads.

4.1 Solution of the Lagrange Equation

As a first step towards the derivation of the electron optical properties of the cylindrical field in the presence of space charge, we solve the equation of motion without space charge. We shall thus recover the basic optical properties of the cylindrical field, as already discussed in Sect. 3.1. The solution for the trajectories without space charge will subsequently be used to calculate the space charge. The Lagrangian in cylinder coordinates reads

$$L = \frac{m}{2}(\dot{r}^2 + r^2\dot{\theta}^2) + c\ln r \ , \tag{4.1}$$

where r and θ are the radial and angular coordinates, respectively, m is the electron mass, and c is a constant, which will be expressed in terms of the pass energy shortly. The equation of motion for the angular coordinate θ

$$\frac{d}{dt}\frac{dL}{\partial\dot{\theta}} = \frac{\partial L}{\partial\theta} \tag{4.2}$$

tells us that

$$\frac{d}{dt}mr^2\dot{\theta} = 0 \tag{4.3}$$

and $r^2\dot{\theta}$ is therefore a constant of motion. The radial equation reads

$$\frac{d}{dt}\frac{\partial L}{\partial\dot{r}} = \frac{\partial L}{\partial r} \ , \tag{4.4}$$

$$m\ddot{r} = mr\dot{\theta}^2 + \frac{c}{r}$$ (4.5)

For an electron travelling on a circle with radius r_0 ($\dot{r} \equiv 0$), we see that

$$c = -mr_0^2\dot{\theta}_0^2 = -2E_0 .$$ (4.6)

Inserting (4.6) in (4.5) yields the equation of motion

$$\ddot{r} - r\dot{\theta}^2 + \frac{2E_0}{mr} = 0 .$$ (4.7)

Since we are interested in the trajectories in the form $r(\theta)$ rather than as $r(t)$, we eliminate the time from (4.7) with the aid of the identity

$$\ddot{r} \equiv r''\dot{\theta}^2 + r'\ddot{\theta} ,$$ (4.8)

where r' denotes the derivative with respect to θ. From (4.3) one obtains

$$2r\dot{r}\dot{\theta} + r^2\ddot{\theta} = 0 \quad \text{and therefore}$$ (4.9)

$$\ddot{r} = r''\dot{\theta}^2 - \frac{2r'^2\dot{\theta}^2}{r} .$$ (4.10)

Inserting (4.10) into (4.7) yields the equation for the trajectories

$$r'' - 2\frac{r'^2}{r} - r + \frac{2E_0}{mr\dot{\theta}^2} = 0 .$$ (4.11)

The angular velocity in the equation may be replaced by the angular velocity at the entrance slit by recalling again (4.3):

$$\dot{\theta}^2(r, \theta) = \frac{r_0^4}{r^4}\dot{\theta}^2(r_0, 0) .$$ (4.12)

We allow for deviations ΔE from the nominal pass energy E_0 and also for trajectories traversing the entrance slit with an angle α_1 with respect to the tangent to the circle of radius r_0. The energy at the entrance position is then

$$\Delta E + E_0 = \frac{m}{2}(\dot{r}^2 + r_0^2\dot{\theta}^2(r_0, 0)) = \frac{m}{2}r_0^2\dot{\theta}^2(r_0, 0)(1 + \tan^2\alpha_1) .$$ (4.13)

The last term in (4.11) may therefore be replaced by

$$\frac{2E_0}{mr\dot{\theta}^2} = \frac{r^3}{r_0^2}\left(1 + \alpha_1^2 - \frac{\Delta E}{E_0}\right)$$ (4.14)

when terms up to second order in α_1 and first order in $\Delta E/E_0$ are retained. It is useful to introduce a reduced radial coordinate ϱ by

$$r = r_0(1 + \varrho) .$$ (4.15)

The final equation for the trajectories up to second order in α_1 is then

$$\varrho'' + 2\varrho - \frac{\Delta E}{E_0} = 2\varrho'^2 - 3\varrho^2 - \alpha_1^2 . \tag{4.16}$$

For a first-order solution we need to retain merely the terms on the left hand side of (4.16). Such a solution will clearly be of the form

$$\varrho = a_1 \sin \omega\theta + a_2 \cos \omega\theta + a_3 \tag{4.17}$$

and with the initial condition

$$\varrho(0) = \varrho_1 , \tag{4.18}$$

$$\varrho'(0) = \alpha_1 , \tag{4.19}$$

we obtain

$$\varrho = \frac{1}{\sqrt{2}} \alpha_1 \sin \sqrt{2}\,\theta + \varrho_1 \cos \sqrt{2}\,\theta + \frac{1}{2} \frac{\Delta E}{E_0} (1 - \cos \sqrt{2}\,\theta) . \tag{4.20}$$

From (4.20) we recover condition (3.3) for the first-order focus, when $d\varrho/d\alpha_1 = 0$, which occurs when $\sin \sqrt{2}\theta_f = 0$. The focus is therefore at

$$\theta_f = \frac{\pi}{\sqrt{2}} = 127.28° . \tag{4.21}$$

The solution including second-order terms in α_1 is obtained when the first-order solution (4.20) is inserted into the right hand side of (4.16) and the differential equation (4.16) is then treated as an inhomogeneous differential equation. After some algebra we find

$$\varrho'' + 2\varrho - \frac{\Delta E}{E} = \frac{7}{4}\alpha_1^2 \cos 2\sqrt{2}\,\theta - \frac{3}{4}\alpha_1^2 , \tag{4.22}$$

for which a particular solution will have the form

$$\varrho_p = a_1 \cos \sqrt{2}\,\theta + a_2 \cos 2\sqrt{2}\,\theta + a_3 . \tag{4.23}$$

The final trajectory equation up to the second order in α_1 then reads

$$\varrho = \frac{\alpha_1}{\sqrt{2}}\theta + \alpha_1^2 \left(\frac{2}{3} \cos \sqrt{2}\,\theta - \frac{7}{24} \cos 2\sqrt{2}\,\theta - \frac{3}{8} \right)$$
$$+ \frac{\Delta E}{2E_0}(1 - \cos \sqrt{2}\,\theta) + \varrho_1 \cos \sqrt{2}\,\theta . \tag{4.24}$$

At the first-order focal point $\theta_f = \pi/\sqrt{2}$ we find for the reduced radial coordinate

$$\varrho_2 = -\varrho_1 + \frac{\Delta E}{E_0} - \frac{4}{3}\alpha_1^2 , \tag{4.25}$$

which is (3.4) in the reduced units ϱ. Up to this point our treatment of the problem has followed the established path [4.1]. Repeating the calculation here was necessary however for the next step, namely, calculation of the space charge and the elctric field caused by the space charge.

4.2 Analytical First-Order Solutions for the Space Charge Problem

We assume that a beam of electrons is incident on a cylindrical deflector. The distribution of angles α_1, energies ΔE and entrance positions ϱ_1 in the beam is assumed to be of the following form

$$f(\alpha_1, \Delta E, \varrho_1) = f_\alpha(\alpha_1) f_E(\Delta E) f_\varrho(\varrho_1) \quad \text{with} \tag{4.26}$$

$$f_\alpha(\alpha_1) = \begin{cases} 1, & |\alpha_1| \le \alpha_m, \\ 0, & \text{otherwise}, \end{cases}$$

$$f_E(\Delta E) = \begin{cases} 1, & |\Delta E| \le \Delta E_{in}/2, \\ 0, & \text{otherwise}, \end{cases}$$

$$f_\varrho(\varrho_1) = \begin{cases} 1, & \varrho_1 \le s_1/2r_0, \\ 0, & \text{otherwise}, \end{cases}$$

where α_m is the angular aperture of the input beam, ΔE_{in} is the total width of the energy distribution and s_1 is the width of the entrance slit. By virtue of the fact that the first-order solution for the trajectories (4.20) is a linear function of α_1, ΔE and ϱ_1, the space charge in a cylindrical deflector subject to these conditions for the entrance beam is symmetric around the central radius r_0. As the beam enters the cylindrical deflector, the space charge spreads out as the bundle of trajectories broadens according to the distribution in α_1, ΔE and ϱ_1. A simple analytical solution for the space charge is found in three limits

$$1) \quad \frac{\alpha_m}{\sqrt{2}} \gg \frac{\Delta E_{in}}{2E_0}, \quad \frac{s_1}{2r_0}, \tag{4.27}$$

$$2) \quad \frac{\Delta E_{in}}{2E_0} \gg \frac{\alpha_m}{\sqrt{2}}, \quad \frac{s_1}{2r_0}, \tag{4.28}$$

$$3) \quad \frac{s_1}{2r_0} \gg \frac{\Delta E_{in}}{2E_0}, \quad \frac{\alpha_m}{\sqrt{2}}. \tag{4.29}$$

The bundle of trajectories which produce the space charge is then described to first order in α_1 by

$$\varrho_{sc} = \frac{\alpha_1}{\sqrt{2}} \sin \sqrt{2}\,\theta, \tag{4.30}$$

$$\varrho_{sc} = \frac{\Delta E}{2E_0}(1 - \cos \sqrt{2}\,\theta), \quad \text{and} \tag{4.31}$$

$$\varrho_{sc} = \varrho_1 \cos \sqrt{2}\,\theta, \tag{4.32}$$

respectively. Case 1 is of the highest practical importance and is therefore discussed first. Because of the linear spreading of the space charge in α_1 the space charge $en(r, \theta)$ at a particual angle θ is

48

$$en(r, \theta) = \frac{c}{r(\theta)\dot\theta(r, \theta)(\alpha_m/\sqrt{2})\sin\sqrt{2}\,\theta} \equiv \frac{c}{r(\theta)\dot\theta(r, \theta)\varrho_m(\theta)} \; . \tag{4.33}$$

The tangential velocity $r\dot\theta$ enters because the space charge is inversely proportional to the velocity. The constant c is determined by the input current I_{in}.

$$I_{in} = eh \int_{r_0(1-\varrho_m(\theta))}^{r_0(1+\varrho_m(\theta))} r(\theta)\dot\theta(r, \theta)n(r, \theta)dr = 2hcr_0 \; . \tag{4.34}$$

Here again h is the height of the entrance slit, $v_\theta = r\dot\theta$ is the velocity of the electrons and $\varrho_m(\theta)$ is the maximum value of the (reduced) radial coordinate as defined by (4.33). Integration of (4.34) yields the space charge, which, except for higher order terms, depends on θ only, in our model:

$$en(\theta) = \frac{I_{in}r}{2v_0 hr_0^2 \cos\alpha} \frac{1}{\varrho_m(\theta)} \; . \tag{4.35}$$

Here we have replaced v_θ by $(r_0/r)v_0 \cos\alpha$ using (4.12) and (4.13). The velocity v_0 is the velocity of electrons travelling along the central radius r_0. The space charge gives rise to an electric field vector which has longitudinal and transverse (radial) components. The longitudinal component is small however and has a minory effect on the trajectories anyway. By neglecting the longitudinal component we have for the radial component of the electric field vector

$$\varepsilon_0 \frac{1}{r} \frac{\partial}{\partial r} r\mathcal{E}_r = en(\theta) \; , \tag{4.36}$$

$$r\mathcal{E}_r = \frac{I_{in}}{2\varepsilon_0 r_0^2 hv_0 \cos\alpha} \frac{1}{\varrho_m(\theta)} \frac{1}{2} \int_{r_0(1-\varrho)}^{r_0(1+\varrho)} r^2 dr \; , \tag{4.37}$$

$$\mathcal{E}_r = \frac{I_{in}}{2\varepsilon_0 v_0 h \cos\alpha} \frac{\varrho(\theta)}{\varrho_m(\theta)} \frac{1+\frac{1}{3}\varrho^2(\theta)}{1+\varrho(\theta)} \; . \tag{4.38}$$

With this additional space charge field the differential equation for the trajectories (4.16) is replaced by

$$\varrho'' + 2\varrho = C_R \frac{1}{\cos^3\alpha} \frac{\varrho}{\varrho_m}(1+\varrho)^3 \left(1+\frac{1}{3}\varrho^2\right)$$
$$+ \frac{\Delta E}{E} + 2\varrho'^2 - 3\varrho^2 - \alpha^2 - 2\varrho'^2\varrho - 3\varrho\alpha^2 - \varrho^3 - \frac{2}{3} \; , \tag{4.39}$$

where we have introduced a space charge coefficient C_R

$$C_R = \frac{eI_{in}r_0}{2\varepsilon_0 mv_0^3 h} = \frac{I_{in}r_0}{4hkE_0^{3/2}} \quad \text{with} \tag{4.40}$$

$$k = \frac{\varepsilon_0}{e}\sqrt{\frac{2}{m}} = 5.25\,\text{mA/eV}^{3/2} \; , \tag{4.41}$$

49

the universal space charge constant as already introduced earlier. The differential equation (4.39) is correct up to all orders in α_1, save for the space charge term and up to first order in ΔE and ϱ_1. We have written the equation such that it has the same form also in the limits 2 and 3 (4.28, 29). For the moment we are interested in the first-order effects of the space charge. We therefore disregard higher order terms in (4.39) in the following discussion and shall return to this issue later in connection with the numerical analysis. The simplified form of the trajectory equation then reads

$$\varrho'' + 2\varrho = C_R \frac{\varrho}{\varrho_m} + \frac{\Delta E}{E} \, . \tag{4.42}$$

For case 1 the essential space charge term is explicitly written

$$\frac{\varrho(\theta)}{\varrho_m(\theta)} = \frac{(\alpha_1/\sqrt{2}) \sin \sqrt{2}\theta + \frac{1}{2}(\Delta E/E_0)(1 - \cos \sqrt{2}\theta) + \varrho_1 \cos \sqrt{2}\theta}{(\alpha_m/\sqrt{2}) \sin \sqrt{2}\,\theta} \, . \tag{4.43}$$

The first part of the numerator gives rise to an increase of the focal length as we shall see shortly, the second produces a current-dependent energy dispersion, and the third a current-dependent magnification. An analytical solution for the latter two effects can be obtained only in the limiting cases 2 and 3 where the space charge is determined by the energy spread of the incoming beam and the slit width, respectively. Here we proceed further by assuming that the incoming beam has no energy spread at all and that the slit width is exactly zero.

We then have for the space charge term the simple form $C_R \alpha_1/\alpha_m$ and the solution for the trajectories is

$$\varrho = \frac{\alpha_1}{\sqrt{2}} \sin \sqrt{2}\,\theta + \frac{1}{2} C_R \frac{\alpha_1}{\alpha_m}(1 - \cos \sqrt{2}\,\theta) \, . \tag{4.44}$$

The first-order focus occurs where $d\varrho/d\alpha_1$ is equal to zero. If we denote the increase of the focal length due to space charge by $\Delta\theta_{sc}$

$$\Delta\theta_{sc} = \theta_f - \frac{\pi}{\sqrt{2}} \tag{4.45}$$

we find

$$\Delta\theta_{sc} = \sqrt{2} \arctan \frac{C_R}{\alpha_m \sqrt{2}} \quad \text{or} \tag{4.46}$$

$$\Delta\theta_{sc} \approx \frac{C_R}{\alpha_m} = I_{in} \frac{r_0}{4hkE_0^{3/2}} \alpha_m \, . \tag{4.47}$$

We see that an important consequence of the space charge is that the first-order focus is shifted to larger deflecting angles. In other words, if a deflecting angle of 127° is used, the electron optical properties of such a device would deteriorate as soon as the deflector is subject to a current load. This fundamental first-order property, although briefly mentioned in [4.2], has not yet been properly recognized. The obvious consequence for practical designs would be to have monochromators with enlarged deflecting angles. In previously published designs however, the deflecting angle of monochromators has apparently always been calculated for a cylindrical deflector without current load. This raises the interesting question of how earlier spectrometer designs actually worked at all, with a reasonable performance. The answer is that the current-dependent first-order term in (4.44) can be balanced against the second-order terms when the distribution of angles of incidence is not centred around $\alpha_{10} = 0$, as was assumed so far, but around some finite current-dependent angle $\alpha_{10}(I_{in})$. Taking (4.44) at the focal angle $\theta_f = \pi/\sqrt{2}$ and adding the second-order terms (without space charge) from (4.24) yields

$$\varrho_2 = C_R \frac{\alpha_1}{\alpha_m} - \frac{4}{3}\alpha_1^2 , \tag{4.48}$$

which is a shifted parabola. First-order focusing is achieved when $d\varrho/d\alpha_1 = 0$, which occurs when the beam is centred around

$$\alpha_{10}(I_{in}) = \frac{3}{8}\frac{C_R}{\alpha_m} . \tag{4.49}$$

As spectrometers are always tuned for optimum transmission and the lens system has some degree of flexibility, which permits the monochromator to be fed with a nonzero mean entrance angle, finite currents can be passed through the monochromator without deterioration of the resolution. As a matter of fact, in our laboratory the lens systems of the cathode were always divided into two segments to allow the feed beam to be deflected. For older spectrometers without enlarged deflection angles in the monochromators, it was consistently noticed that the cathode lens system operated with quite appreciable sideways deflecting potentials on the lenses for optimum performance. Unfortunately we, and presumably other researchers, attributed this lens operation to some spurious effect rather going through the straightforward analysis presented above. This analysis tells us also that operating with deflected beams is by no means equivalent to enlarging the deflection angle θ_f. Equations (4.47) and (4.49) show that an extension of θ_f by 20° is equivalent to $\alpha_{10} = 7.5°$. While the former is quite feasible, α_{10} is limited to a few degrees by the angular aberrations of the lens system, which follows the monochromator.

We now briefly discuss the second and the third limits, (4.28) and (4.29), where an analytical solution is available. If

$$\frac{\Delta E_{in}}{2E_0} \gg \frac{\alpha_m}{\sqrt{2}} , \quad \frac{s_1}{2r_0} , \tag{4.50}$$

the first-order equation for the trajectories is

$$\varrho'' + 2\varrho = \frac{\Delta E}{E_0}(1 + C_R\frac{E_0}{\Delta E_{in}}) . \tag{4.51}$$

51

It follows immediately that the energy dispersion is

$$E_0 \frac{d\varrho}{dE} = 1 + C_R \frac{E_0}{\Delta E_{in}} = 1 + \frac{I_{in} r_0}{4 h k E_0^{1/2} \Delta E_{in}} . \tag{4.52}$$

Thus the energy dispersion increases with the current, which means that monochromators carrying space charge may actually have a higher resolution than normal analysers, when they are properly designed. Finally we have the equation for the limit of a large slit width,

$$\varrho'' + 2\varrho = 2r_0 C_R \frac{\varrho_1}{s} , \tag{4.53}$$

from which the magnification is easily calculated to be

$$C_y = - \left(1 - 2C_R \frac{r_0}{s} \right) . \tag{4.54}$$

This last result is however of little practical use since monochromators work best with small slit widths. For such systems, the modulus of the magnification increases nearly linearly with the current rather than becoming smaller as suggested by (4.54). Furthermore, for monochromators with equipotential apertures at the entrance and exit positions, the magnification remains near $C_y = -1$ even at higher currents, unless retarding deflectors are used.

This short and straightforward treatment of the space-charge-induced modifications of the first-order imaging properties of a cylindrical deflector contains the nucleus of the design principles upon which optimised space-charge-limited monochromators should be based. Some additional information is needed for the extension of these considerations into the three-dimensional world, and furthermore, information is also needed on the effect of the space charge on the higher order angular aberration and on the effect of equipotential apertures. Before we move on to such matters we pause for a moment and consider the effect of space charge on another electrostatic deflector that has been used frequently, the spherical deflector.

4.3 Space Charge in a Spherical Deflector

We now begin with the Lagrangian for the ideal *spherical* field

$$L = \frac{m}{2} \left[\dot{r}^2 + r^2 (\dot{\theta}^2 + \sin^2 \theta \dot{\varphi}^2) \right] + \frac{c}{r} . \tag{4.55}$$

We are interested in solutions for which the electrons travel in the tangential plane along meridians of the sphere, i.e. when $\dot{\varphi} = 0$. Proceeding as before, we arrive at the trajectory equation

$$\varrho'' + \varrho \frac{\Delta E}{E_0} = 2\varrho'^2 - \varrho^2 - \alpha^2 , \tag{4.56}$$

where we have included terms up to second-order as in (4.16). Electrons travelling along meridians (with $\varrho'' \equiv 0$) intersect in the two poles. This means that one has automatically a focusing property with respect to the azimuthal angle φ. The trajectories in the radial plane up to second order are calculated from (4.56) following the procedure described in Sect. 4.1:

$$\varrho = \alpha \sin\theta + \varrho_1 \cos\theta + \frac{\Delta E}{E}(1 - \cos\theta) + \alpha^2 \left(\cos\theta - \frac{1}{2}\cos 2\theta - \frac{1}{2} \right) . \quad (4.57)$$

We note that the energy dispersion is twice as high as for the cylindrical deflector. The first-order focus in the radial plane is also at 180°, which shows that the spherical deflector has stigmatic focusing. The angular aberration there is $-2\alpha^2$, which is larger than for the cylindrical deflector.

The angular aberration has an adverse effect on the resolution as it does for the cylindrical deflector, and for optimum performance the maximum angle α_m should be limited to a few degrees. No such limitation is needed in the other direction, as all electrons travelling along a meridian are perfectly focused. There the limitation is usually due to the lens system either before or after the spherical deflector. The unusual focusing properties of the spherical deflector have been used to determine energy and emission angle of photoexcited electrons simultaneously [4.3]. Here we concentrate on the properties of spherical deflectors when used as monochromators. Spherical deflectors may be used with circular apertures but also with slit apertures, preferably in the form of slits shaped according to the mean radius r_0. The optimum extension of the slit in the azimuthal plane is again subject to considerations similar to those for the cylindrical deflector. The width of the slits s is directly related to the base width of the energy distribution.

$$\frac{\Delta E_B}{E_0} = \frac{s}{2r_0} + \alpha_m^2 . \quad (4.58)$$

In order to study the effect of space charge on the trajectories, we make the simplifying assumption that the space-charge-induced electric field has only a radial component and that the trajectories for the electrons are the first-order trajectories of (4.57). We discuss the situation where

$$\alpha_m \gg \frac{\Delta E_{in}}{E} , \quad \varrho_1 . \quad (4.59)$$

For the first-order trajectory equation, we then have

$$\varrho'' + \varrho = C_R \frac{\alpha}{\alpha_m} \quad (4.60)$$

where C_R denotes the same space charge coefficient as before. The solution is

$$\varrho = \alpha \sin\theta + C_R \frac{\alpha}{\alpha_m}(1 - \cos\theta) . \quad (4.61)$$

The current-dependent term linear in α requires an extension of the deflection angle in order to make the linear α-term vanish. The extension here is

$$\Delta\theta_{sc} = \frac{2C_R}{\alpha_m} , \tag{4.62}$$

i.e. twice as high as for the cylindrical deflector. Since the first-order focus is already at 180°, the need for an extension of the deflection angle considerably beyond that would make a spherical monochromator adjusted for space charge somewhat inconvenient to use. More importantly, however, the spherical deflector extended beyond 180° would *lose the property of stigmatic focusing* since the meridional focus is not (or rather less) affected by space charge. So far, no spherical analyser with an extended deflecting angle has been constructed. Spherical monochromators in existing designs presumably work with a nontangential feed beam. The angular aberration term with space charge at $\theta = \pi$ is

$$\varrho_1 = 2C_R \frac{\alpha_1}{\alpha_m} - 2\alpha_1^2 , \tag{4.63}$$

which provides for first-order focusing ($d\varrho_1/d\alpha_1 = 0$) when the bundle of trajectories feeding the monochromator is centred around

$$\alpha_{10} = \frac{1}{2} \frac{C_R}{\alpha_m} . \tag{4.64}$$

This offset is higher than for the cylindrical deflector by a factor of 4/3 when referred to the same feed current, radius, slit height and energy. So far, no deliberate attempt has been reported to adjust the lens systems used to feed the monochromator or the transport lens to the sample in such a way as to match the angular offset required by the monochromator. If such lenses were used, one would generate probably about the same monochromatic currents with spherical deflectors as with cylindrical deflectors operating with an offset angle α_{10} when slits rather than round apertures are used. The stigmatic focusing properties of the spherical deflector do however, also require the transformation of a stigmatic image of the exit slit of the device on the sample and on the entrance slit of the analyser in order to achieve optimum transmission. Electrostatic lenses with their rather high aberration, especially when operated with large acceleration-retardation factors, tend to form rather poor images of slits. The same problem does not arise with cylindrical deflectors as the lens system there is *not* required to form a stigmatic image of the exit slit. We shall discuss this matter at length later. Here we merely note that the combination of electron optical properties of monochromators and lenses gives the cylindrical deflector some advantages over the spherical deflector, which adds to the fact that cylindrical deflectors are easier to fabricate. We therefore concentrate on the cylindrical deflector.

4.4 Numerical Calculation of Space Charge Effects

We now turn to the numerical analysis of the cylindrical deflector with space charge. We begin with the ideal cylindrical field since there we have already derived analytical expressions for the focal length, the energy dispersion, and the magnification. Comparison of the numerical results with the analytical expressions is useful for testing the accuracy and convergence of the computer codes. One may also derive interpolation formulas to bridge the gap between the limits for which we have found analytical solutions. In particular, the energy dispersion in the limit $\alpha \gg \Delta E_{in}/E_0$ and the optimum focal length in the limit $\Delta E_{in}/E_0 \gg \alpha_m$ is of importance, as is the magnification. Finally, one needs information about the angular aberrations in the presence of space charge, as they ultimately determine the resolution of the monochromator. Before we present the results of the fully numerical analysis, we perform the direct numerical integration of the equation of motion (4.39) as a first step. The example of the trajectories and the radial positions at the deflection angle of 139.9° is shown in Fig. 4.1. The input energy distribution is assumed to be of zero width. The input

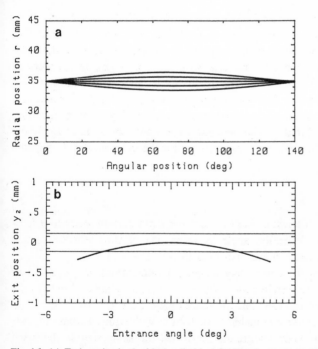

Fig. 4.1. (a) Trajectories in the ideal cylindrical field in the presence of space charge calculated by direct numerical integration of (4.39). The energy width of the input beam is assumed to be zero. The first-order focus at 139.9° is obtained when the input current is 4.2×10^{-8}, when $E_0 = 1\,\mathrm{eV}$, $r_0 = 35\,\mathrm{mm}$ and $\alpha_m = \pm 3°$. (b) The radial exit position as a function of the input angle α is a parabola as for the ideal cylindrical field without space charge. Within the limits of error the coefficient $C_{\alpha\alpha}$ is the same as without space charge

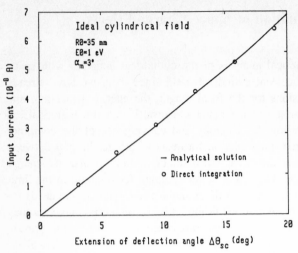

Fig. 4.2. Comparison of the input current leading to a first-order focus as a function of the deflection angle according to the first-order analytical solution (4.46) and as obtained by numerical integration of (4.39)

current for which the first-order focus occurs at $\theta_f = 139.9°$, is 4.2×10^{-8} A, when the energy $E_0 = 1\,\text{eV}$, the radius $r_0 = 35\,\text{mm}$, and the angular aperture of the input beam $\alpha_m = \pm 3°$. For the purposes of integration of the differential equation (4.39), we have replaced the term $\varrho(\theta)/\varrho_m(\theta)$ by α/α_m in the spirit of the limit $\Delta E_{in} = 0$, $\varrho_1 = 0$ (4.43). Unlike the first-order analytical solution (4.44) where all higher order terms were neglected, we have performed the numerical integration of (4.39) with all higher order terms included. Figure 4.1b nevertheless shows that the angular aberration remains second order in α_1 and the angular aberration coefficient $C_{\alpha\alpha}$ is nearly the same as for the cylindrical field without space charge, namely

$$C_{\alpha\alpha} \approx -\tfrac{4}{3}r_0 .$$ (4.65)

This is a quite remarkable result. Furthermore the input current leads to a first-order focus at a given deflection angle as a function of this deflecting angle comes out just as calculated in the first-order theory (Fig. 4.2).

We now embark on the completely numerical calculations of the effect of space charge on the electron trajectories. We take the ideal cylindrical field as the basis. The procedure used to calculate the space charge potential, to be described in the following, is likewise applicable to deflectors with equipotential entrance and exit apertures. The basic integration mesh had the same size as that used earlier, namely 100×200 in the radial and tangential directions, respectively. The 200 units in the tangential direction correspond to a deflection angle of $\theta = \pi/\sqrt{2} = 127.3°$. The array was extended beyond 200 when larger deflecting angles were considered. The space charge was calculated by defining an integer array $R(I, J)$ on the 100×200 mesh. After each step of the numerical integration

56

of a trajectory the integer array was raised by one unit for a particular I and J when the instantaneous position of the electron corresponded to the domain $I - 0.5$, $J - 0.5$; $I + 0.5$, $J - 0.5$; $I - 0.5$, $J + 0.5$; $I + 0.5$, $J + 0.5$. The space charge integer array was typically filled with 400–1000 trajectories with initial conditions randomly distributed over the slit width, the width of the energy distribution and the angular aperture. In order to avoid systematic fluctuations in the space charge array along the tangential coordinate we also randomised the integration time unit within a factor of two and the starting position with respect to the tangential mesh size. The residual noise in the space charge field was reduced by digital averaging over nearest neighbors in the integer array. We note that digital averaging along the radial coordinate essentially simulates space charge produced by a bundle of trajectories with a smooth distribution in angles rather than with a sharp cut off at the angle α_m. If we reconsider the derivation of the differential equation (4.39) and the solution of this equation to first order, it becomes evident that the higher order angular aberration terms induced by the space charge potential are influenced by the space charge distribution along the radial coordinate, and thus also by the shape of the angular distribution of the feed beam. Since the shape of the angular distribution in real systems is not well known and may also vary, the angular aberration terms which result from the numerical calculation have to be considered with some reservation, when space charge is involved.

The integer field R representing the space charge is converted to a field which represents the space charge ϱ by

$$\frac{\varrho(I, J)}{\varepsilon_0} = \frac{R(I, J) I_{\text{input}}}{\bar{v}_0 h \Delta r \sum_J R(I, J)} , \tag{4.66}$$

where I_{input} is the input current, $\bar{v}_0$ is the average initial velocity, h the slit height, and Δr the length of the basic mesh along the radial coordinate. When the width of the energy distribution in the feed beam is not small compared with the pass energy, it is important to take $\bar{v}_0$ as the mean velocity in the energy distribution and not as the velocity of the mean energy. With this space charge, the Poisson equation was solved as described in Sect. 3.2. The boundary conditions were such that the space charge induced potential is zero at the outer and inner deflection plates. For the terminated deflector, one has the additional boundary condition that the potential should likewise be zero at the entrance and exit apertures, while for the ideal cylindrical field one has periodic boundary conditions. As a result of these different boundary conditions, the equipotential contours of the space-charge-induced potential have a quite different appearence, Fig. 4.3. For the ideal cylindrical field, the tangential electric field components are vanishingly small, while for the terminated deflector, appreciable space-charge-induced tangential field components are present near the entrance and exit apertures.

The convergence of the Laplace algorithm is substantially speeded up when one sets out from a potential array close to the final converged result. A useful approach to this end is to use the potential of a space charge distribution in the form of a δ-function in the r direction situated on that particular radial

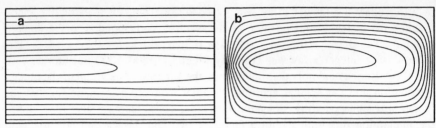

Fig. 4.3. Contours of a space charge potential (a) for the ideal cylindrical field and (b) for a cylindrical deflector terminated by equipotential plates. The energy width of the feed beam is 0.3 eV, the pass energy 1 eV and $\alpha_m = \pm 3°$. The radii of the inner and outer deflection plates are 25 and 45 mm, respectively

coordinate that represents the radial centre of the space charge array for each angular coordinate. After some algebra, this potential is found to be given by

$$V_1(r,\theta) = \tilde{Q}(\theta) r_{\mathrm{sc}} \frac{\ln(R_2/r_{\mathrm{sc}})}{\ln(R_2/R_1)} \ln\left(\frac{r}{R_1}\right) , \quad R_1 \leq r \leq r_{\mathrm{sc}} ,$$

$$V_2(r,\theta) = \tilde{Q}(\theta) r_{\mathrm{sc}} \frac{\ln(R_1/r_{\mathrm{sc}})}{\ln(R_2/R_1)} \ln\left(\frac{r}{R_2}\right) , \quad r_{\mathrm{sc}} < r \leq R_2 .$$

(4.67)

Here r_{sc} denotes the radial position of the δ-function, R_1 and R_2 have the usual meaning of the radii of the inner and outer deflection plates, respectively, and $\tilde{Q}$ is the weight of the δ-function obtained by integration of the space charge density along the radial coordinate

$$\tilde{Q}(\theta) = \int en(r,\theta) dr = \sum_J en(J,\theta) \Delta r .$$

(4.68)

When (4.67) is used as the starting potential, the repeated application of the Laplace algorithm (2.6) merely reshapes the potential in the area where the space charge occurs. The discontinuity in the second derivative of the potential disappears and the potential becomes a smooth curve, which is approximately a parabola. Sufficient convergence is achieved with a few hundred iterations of the Laplace algorithm. The space charge and the converged field are illustrated in Fig. 4.4.

Once the space charge potential has been calculated, the result is used to integrate the trajectories in the presence of space charge; the space charge potential is multiplied by a factor representing a particular input current I_{in} relative to the standard current I_0 used for calculating the space charge potential (4.65). Subsequently, the space charge potential is added to the potential of the bare deflector, where we again make use of the superposition principle. For a particular choice of input current, one thus obtains a set of trajectories. A small subroutine then serves to find the input current that renders the first-order focus at the particular deflecting angle which one wishes to investigate. Rather than search for the current where the linear term in the expansion of the radial coordinate y_2 as a

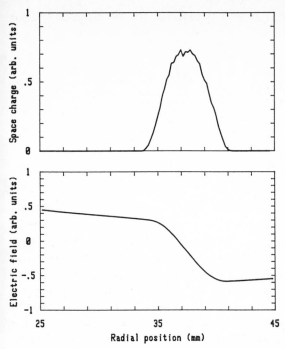

Fig. 4.4. Space charge and the space-charge-generated electric field in the centre of a cylindrical deflector. The space charge refers to an energy width of the feed beam of 0.3 eV, a pass energy of 1 eV and $\alpha_m = \pm 3°$

function of α_1

$$y_2(\alpha_1) = C_\alpha \alpha_1 + C_{\alpha\alpha} \alpha_1^2 + C_{\alpha\alpha\alpha} \alpha_1^3 \ldots \tag{4.69}$$

vanishes, one may also seek the current that provides a minimum spread in y_2 in the α-regime of interest. The latter method yields more realistic results when the third–order coefficient $C_{\alpha\alpha\alpha}$ contributes significantly.

In the first-order analytical treatment as well as in the direct integration of the differential equation (4.39), it was assumed that the space charge distribution behaves as if there were no higher order angular aberrations in the trajectories representing the space charge. On the other hand, by making the substitution

$$\frac{\varrho(\theta)}{\varrho_m(\theta)} = \frac{\alpha_1}{\alpha_m} \tag{4.70}$$

the increase in the focal length due to the space charge is approximately accounted for. When the space charge is calculated numerically one uses as a first step the ideal cylindrical field to calculate the trajectories which have their focus at 127.3°. Subsequently the space charge and the space charge potential are determined. In a second iteration one again calculates the trajectories, optimises the current as described before and repeats the calculation of the space charge and space charge potential and so forth. Clearly the convergence of the procedure depends

59

Table 4.1. Effect of the space charge on the first-order properties of the ideal cylindrical field after the first, second, and third iterations

$\Delta\theta_{sc}$		9.4	12.6	15.8	19.0
	1	2.73	3.56	4.17	4.85
$I_{in}\ [10^{-8}\,\text{Å}]$	2	3.38	4.38	5.48	6.06
	3	3.38	4.30	5.29	6.51
	1	1.28	1.35	1.46	1.55
$E_0\delta\varrho/\delta E$	2	1.23	1.28	1.36	1.46
	3	1.16	1.27	1.34	1.43
	1	1.45	1.58	1.74	1.84
$-C_y$	2	1.41	1.53	1.68	1.85
	3	1.42	1.53	1.67	1.84

on the increase of the focal length or the current for which one wishes to have the numerical result. For $\Delta\theta_{sc}$ up to 20° we have found three iterations to be sufficient for converged results (Table 4.1). The convergence test, when carried to more iterations, also serves to estimate the noise in the results, which is a consequence of the finite number of iterations in the Laplace algorithm and of the finite number of trajectories used to calculate the space charge array. For the purpose of practical design the first iteration suffices, in particular when the calculation is performed with the 3D algorithm (Sect. 5.1).

The results for the optimum input current, the dispersion, and the magnification as a function of the extension of the deflection angle $\Delta\theta_{sc}$ after three iterations of the procedure described above are shown in Fig. 4.5. The input energy distribution is assumed to have zero width, which makes the calculated input current for a particular extension of the deflection angle a lower limit. The width of the angular distribution was assumed to be $\alpha_m = \pm 3°$ and the slit width $s = 0.3\,\text{mm}$. The choice of α_m is motivated by the considerations that led to (3.36). The radial position of the entrance and exit slits is $r_0 = 35\,\text{mm}$. The radii of the deflecting plates are $R_1 = 25\,\text{mm}$ and $R_2 = 45\,\text{mm}$. The particular choice of the radii of the deflecting plates is however irrelevant here, where we have assumed the basic field to be the ideal cylindrical field. In Fig. 4.5 we have also plotted the input current as calculated from the first-order space charge theory in the limit $\alpha_m \gg \Delta E_{in}/E_0$, s/r_0 using (4.47). The agreement between the model and the converged numerical result is surprisingly good. As we have remarked before, the good agreement is presumably due to the replacement of $\varrho(\theta)/\varrho_m(\theta)$ in (4.39) by α/α_m, which already takes the increase of the focal length introduced by the space charge into account.

The good agreement between computer simulation and the equation for the input current that leads to a first-order focus makes equation (4.47) a good starting point for calculating the monochromatic current produced by a cylindrical deflector. In Fig. 4.5 the other two first-order effects of the space charge, namely, the energy dispersion and the magnification, are also shown as a function of the extension of the deflection angle. We remember that these two quantities could

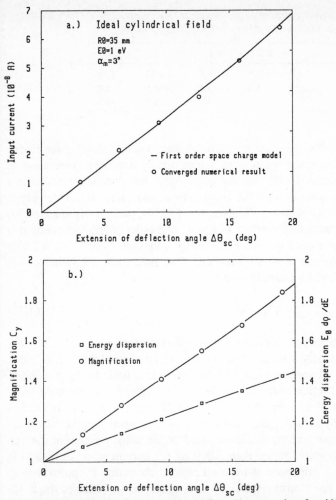

Fig. 4.5a,b. Numerical results concerning the first-order properties of an ideal cylindrical field with a self-consistent space charge field. (a) Input current for first-order focus in α at the deflecting angle $\theta_f = \pi/\sqrt{2} + \Delta\theta_{sc}$ as a function of $\Delta\theta_{sc}$. The full line is calculated from the analytical theory using (4.47). (b) Magnification and energy dispersion as a function of $\Delta\theta_{sc}$

not be calculated analytically in the limit of large α_m although (4.43) made it apparent that the space charge must affect the energy dispersion and the magnification. The increasesd energy dispersion for cylindrical deflectors, when they are properly adjusted to the space charge, is rather welcome as it offers the opportunity for improved resolution. We have also made an attempt to estimate at least the higher order aberration coefficients with space charge. For this purpose we extended the mesh to 200×200 and reduced the gap between the inner and outer deflection plate to 8 mm so that the size of the elementary mesh was 0.04 mm $\times$ 0.389 mm in the radial and tangential directions respectively. The result is shown in Table 4.2. Save for some noise, the second-order aberration

Table 4.2. Approximate angular aberration coefficients for an ideal cylindrical field with space charge. The current load is adjusted to make the first-order coefficient C_α vanish

$\Delta\theta_{sc}$	0	3.1	6.2	9.4	12.6	15.8
$C_{\alpha\alpha}/r_0$	−1.33	−1.41	−1.51	−1.41	−1.33	−1.35
$C_{\alpha\alpha\alpha}/r_0$	~ 0	−2.9	−3.2	−6.9	−10.1	−12.4
$C_{\alpha\alpha\alpha\alpha}/r_0$	~ 0	23	56	44	8.5	40.9

coefficient remains the same as for the ideal cylindrical field without space charge effects. We thus recover the result already obtained by numerical integration of the differential equation (4.39). On the other hand, the third-order coefficient rises approximately linearly with the input current and $\Delta\theta_{sc}$. This latter effect did not emerge from the numerical integration of (4.39), presumably since (4.39) does not take into account the effect of the second-order aberration coefficient on the space charge term appropriately. Finally it appears from Table 4.2 that a fourth-order coefficient is also appreciable. The data there are rather noisy, however. Nevertheless they clearly prove that

$$C_{\alpha\alpha\alpha\alpha}\alpha_m^4 \ll C_{\alpha\alpha\alpha}\alpha_m^3 < C_{\alpha\alpha}\alpha_m^2 . \tag{4.71}$$

The fourth-order aberration therefore has no effect on the resolution. Even the third-order term has only a minor effect on the resolution, unless $\Delta\theta_{sc}$ is much larger. In the next section we shall find that there is a limit to the possibility of enlarging the deflection angle of a deflector. The influence of the third-order coefficient on the resolution therefore remains small. The effect on the resolution may be minimised by adjusting the input current in such a way that the difference between the maximum and the minimum value of the radial position at the exit slit is as small as possible rather than adjusting the current so as to make the linear term vanish. A minimum of the total angular aberration is approximately achieved when the linear term produced by the space charge balances the third-order term at α_{1m}. The remaining total angular aberration is then nearly the same as for the ideal cylindrical field without space charge. Since the input current according to (4.47) is proportional to α_m, the considerations which led us to establish an optimum value for the angular width of the input beam (3.38) as presented in Sect. 3.3 are valid here also.

In the analytical and numerical calculation, we have so far made the unrealistic assumption that the feed beam has zero energy width. We now remove this constraint and calculate the optimum input current as a function of the energy width ΔE_{in} of the feed beam for a particular deflection angle. A typical result is shown in Fig. 4.6. For small ΔE_{in}, the current is nearly constant. As the energy width approaches the boundary of the condition (4.27)

$$\Delta E_{in} = \sqrt{2}\alpha_m E_0 \tag{4.72}$$

the optimum input current begins to rise as the beam spreads over a larger area

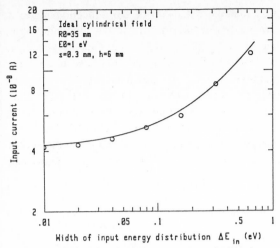

Fig. 4.6. Optimum input current for an ideal cylindrical field with $\Delta\theta_{sc} = 12.6°$ as a function of the energy width of the feed beam. The solid line is a fit with (4.74)

in the monochromator. According to the fundamental trajectory equations (4.20), the spreading should be roughly linear in α_m and ΔE_{in}. For small ΔE_{in}, we had found the analytical expression (4.47)

$$I_{in} = 4hkE_0^{3/2}\alpha_m\frac{\Delta\theta_{sc}}{r_0} \ . \tag{4.73}$$

For larger ΔE_{in} we therefore try replacing the term α_m in the input current by $\alpha_m + \text{const} \times \Delta E_{in}/E_0$, where the constant is chosen to match the result of the computer simulation. This is justified since the dependence of the input current on $\Delta\theta_{sc}$, r_0, and h is of a fundamental nature and does not change. The solid line in Fig. 4.6 represents the fit with the constant equal to 0.0525. The equation for the optimum input current is hence

$$I_{in} = \frac{4hkE_0^{3/2}\Delta\theta_{sc}(\alpha_m + 0.0525\Delta E_{in}/E_0)}{r_0} \ . \tag{4.74}$$

As a final step towards the ultimate goal of this exercise, which is to calculate the monochromatic current produced by the cylindrical deflector, we need to have a quantitative expression for the energy dispersion. This is available in the limit of large ΔE_{in} from (4.52). With (4.74) one has

$$E_0\frac{d\varrho}{dE} = 1 + 0.175\sqrt{\frac{E_0}{\Delta E_{in}}}\Delta\theta_{sc} \ , \quad \Delta E_{in} > \sqrt{2}\alpha_m E_0 \ . \tag{4.75}$$

In the other limit of small ΔE_{in}, we take from Fig. 4.5

$$E_0\frac{d\varrho}{dE} = 1 + 1.3\Delta\theta_{sc} \ , \quad \Delta E_{in} < \sqrt{2}\alpha_m E_0 \ . \tag{4.76}$$

In the next chapter we shall see that the extension of the deflection angle is limited to a few degrees. In an analytical expression for the monochromatic current as a function of the energy width of the output current, the enhanced dispersion may therefore be neglected to first order. We may also neglect the enhanced current-dependent magnification C_y as shown in Fig. 4.5. We shall later see that this enhanced magnification is a property of the ideal cylindrical field. For cylindrical deflectors with equipotential apertures to be studied in the next section, C_y stays near -1, irrespective of the input current and $\Delta\theta_{\rm sc}$.

When we assume that the energy distribution of the feed beam and the energy distribution of the monochromatic current $I_{\rm out}$ have the same (e.g. Gaussian) shape, we have

$$I_{\rm out} = I_{\rm in}\frac{\Delta E_{1/2}}{\Delta E_{\rm in}}\frac{1}{\alpha_{1\rm m}}\int_0^{\alpha_{1m}} T(\alpha_1)d\alpha_1 , \tag{4.77}$$

where $T(\alpha_1)$ is the transmission (3.29). Since the current is $\sim \Delta E_{1/2}^{5/2}$ we take from (3.36) $\alpha_{\rm m\,opt} = \sqrt{s/2r_0}$. The integral over the transmission function is then $7/9$. For small enough $\alpha_{\rm m}$, the monochromatic current is

$$I_{\rm out} = \frac{7}{6}\sqrt{\frac{3}{2}}\frac{\Delta E_{1/2}^{5/2}}{\Delta E_{\rm in}}\frac{h}{s}k\Delta\theta_{\rm sc} . \tag{4.78}$$

In (4.78) we have used the simplified relation for the FWHM of the transmitted beam:

$$\frac{\Delta E_{1/2}}{E_0} = \frac{s}{r_0} + \frac{2}{3}\alpha_{\rm m}^2 . \tag{4.79}$$

We shall see in the next chapter that the presence of the ratio of the slit height over slit width in (4.78) results from the two-dimensionality of the treatment. In three-dimensional calculations, the monochromatic current becomes essentially independent of the slit height. Furthermore the equations for the monochromatic current have to be interpreted with the caveat that $\Delta\theta_{\rm sc}$ is subject to constraints arising from the divergence of the beam in the z-direction and other considerations to be discussed in the next chapter. A quantitative evaluation of the effect of vertical beam divergence on the base width and the transmission requires an extension of the numerical calculations into the third dimension. This extension will be performed with the cylindrical deflector terminated by metal apertures as entrance and exit slits and also equipped with top and bottom shields. Here we merely note that the most effective way of achieving the highest possible monochromatic current is to reduce the energy width of the feed beam by feeding a second monochromator from a pre-monochromator, which is in turn fed by the thermal energy distribution of the cathode. Such two-step monochromatisation is accomplished elegantly with the help of a retarding pre-monochromator, which again operates with equipotential apertures at the entrance and exit slit positions.

5. Electron Optics of Real Cylindrical Deflectors Loaded with High Current

In this chapter monochromators are studied and optimised with the help of numerical simulations of trajectories in the presence of space charge. The advantage of monochromatisation in two (or more) steps is demonstrated. Since it is an advantage to produce a monochromatic beam via a stepwise monochromatisation, we present a numerical analysis of two types of monochromators in the following. In Sect. 5.1 we first discuss the result of computer simulations for the second stage of a two-step monochromator, which operates with the entrance and exit apertures at a potential equal to the average of the potentials of the inner and outer deflection plates. In Sect. 5.2 we present results relevant to pre-monochromators, where either a higher potential is applied to the entrance aperture, or a lower potential to the exit aperture. We have already encountered such retarding monochromators in Sect. 3.2, where the limiting case of a small current load was considered.

5.1 Monochromators

Despite the apparent simplicity of the cylindrical deflector, the number of dimensional parameters and different potentials is quite large. Exploring the full effect of all conceivable design parameters in three-dimensional simulations represents a formidable, if not hopeless, task. The analytical and numerical treatment in Chaps. 5.3 and 4 helps to identify the key parameters, which need further exploration with three-dimensional calculations. In our investigation of the effect of various parameters, we start from a particular reference frame of parameters, which is listed in Table 5.1. We have chosen a pass energy of 0.3 eV to provide a realistic and useful resolution. The choice of the deflection angle was made after many exploratory studies of which the salient results will be discussed in the course of this section. The entrance and exit apertures were assumed to be situated at a distance to the deflecting plates equivalent to a deflection angle of 2.5°, and the gaps between the top and bottom cover plates and the deflection plates were 2 mm. The size of the gaps as well as the precise orientation of the apertures are not critical. The choice of the energy width of the feed beam follows from the following preliminary optimisation.

We have found (4.73) that in the limit of a small energy width of the feed beam, $\Delta E_{in} < \sqrt{2}\alpha_m E_0$, the maximum input current of a monochromator is

Table 5.1. List of design parameters for a reference monochromator

Deflection angle	θ_{sc}	114°
Radius of outer deflection plate	R_2	45 mm
Radius of inner deflection plate	R_1	25 mm
Radius of central path	r_0	35 mm
Total height	H	44 mm
Slit width	s	0.3 mm
Slit height	h	6 mm
Maximum aperture angle	α_{1m}	3°
Maximum aperture angle	β_{1m}	0°
Energy width of feed beam	ΔE_{in}	0.02 eV
Pass energy	E_0	0.3 eV

$$I_{in\,2} = C_{p2}\Delta E_2^{3/2} \,, \tag{5.1}$$

where the index 2 refers to the second monochromator and c_{p2} characterises the design parameters and performance. The same equation, though with a different constant c_{p1}, holds for the first monochromator:

$$I_{in\,1} = c_{p1}\Delta E_1^{3/2} \,. \tag{5.2}$$

Disregarding transmission losses, the output current of the first monochromator is also

$$I_{out\,1} \approx \frac{\Delta E_1}{\Delta E_{cath}}\, I_{in\,1} \,, \tag{5.3}$$

where ΔE_{cath} refers to the width of the energy distribution provided by the cathode emission system. Since the output current of the first monochromator I_{out} should match the maximum input current of the second monochromator (5.1), we have

$$c_{p2}\Delta E_2^{3/2} = \frac{\Delta E_1}{\Delta E_{cath}}\, c_{p1}\Delta E_1^{3/2} \,. \tag{5.4}$$

One may further assume that the constant c_{p1} characterising the retarding monochromator and the constant c_{p2} are about equal. One then finds for the optimum choice of the energy width of the first monochromator

$$\Delta E_1 = \left(\Delta E_{cath}\Delta E_2^{3/2}\right)^{2/5} \,. \tag{5.5}$$

Table 5.2 provides a few numbers for ΔE_1 and ΔE_2 when a typical width of the thermal energy distribution of the cathode of 300 meV is assumed. Thus a choice of 20 meV for the feed beam in the parameter set (Table 5.1) is in line with these considerations. Experimentally, one finds that the monochromatic current is a rather smooth function of the pass energy (i.e. resolution) of the pre-monochromator. The precise value of the pass energy is therefore not critical and typically lies between 3 and 10 times the pass energy of the second monochro-

Table 5.2. Optimum choice for the energy width of a pre-monochromator ΔE_1 as a function of the energy width of the second monochromator ΔE_2 when the energy width of electrons emitted from the cathode is 300 meV

ΔE_2 [meV]	1	2	3	4	5
ΔE_1 [meV]	9.8	14.8	18.9	22.5	25.7

mator. Details are specific to the individual design and will be discussed at length in connection with a particular design in Chap. 8.

One final comment on the set of parameters in Table 5.1 concerns the angular aperture β_{1m}. We have chosen $\beta_{1m} = 0$ because test runs with $\beta_{1m} = 0, 2, 4°$ revealed that although the monochromatic current produced by the deflector not unexpectedly decreases with β_{1m}, the effect is small ($\sim 20\%$) for the quoted values of β_{1m} and a further study of the β_{1m}-dependence would merely lead to the trivial result that one should attempt to keep the angular aperture β_{1m} small. Realistic values of β_{1m} obtainable with cathode emission systems will be discussed in Sect. 6.2.

We have performed the calculation of the potential distribution on a grid of $50 \times 100 \times 24$ units, as we did for the 3D analyser, where the numbers refer to the number of intervals along the r, θ and z coordinates, respectively. Where possible we have compared the results of the trajectory calculations performed on this grid with the two-dimensional calculations on a fine 100×200 grid, where we also used a larger number of integration steps. The main effect of the less accurate calculation is to shift the focus at $\theta_f = 106.8°$ by about 1.9° to a shorter deflection angle. The deflection angles quoted in the following are corrected for this shift of 1.9°. The space charge was calculated with the trajectories of 500 electrons with initial conditions randomly distributed between the boundaries provided by the size of the entrance slit, the energy width of the feed beam, and its angular divergence. An integer field representing the local space charge was created as described in Sect. 4.5, except that now the field is a three-dimensional field $R(I, J, K)$. The normalisation to a space charge field $\varrho(I, J, K)$ is then

$$\frac{\varrho(I, J, K)}{\varepsilon_0} = \frac{R(I, J, K) I_{\text{input}}}{\bar{v}_0 \Delta z \Delta r \sum_{J,K} R(I, J, K)} . \tag{5.6}$$

The index I counts along the θ-coordinate and $\Delta z, \Delta r$ are the dimensions of the grid along the z and r coordinates, respectively. Here again I_{input} is the input current and $\bar{v}_0$ the average velocity in the feed beam. The space charge potential was then calculated by solving the Poisson equation.

As in the 2D calculation, one may use the trajectories calculated with the particular current load that leads to a focus at the extended deflection angle and proceed to calculate the space charge in a second iteration, and so forth. Table 5.3 shows the results for the first-order properties of the cylindrical deflector. One finds that the first iteration is already sufficiently well-converged. We have therefore used only one iteration in the following simulations. We notice also from Table 5.3 that the magnification remains approximately $C_y = -1$, the same

Table 5.3. Effect of the space charge on the first-order properties of a 3D-cylindrical deflector with extension of the focal length of $7°$ ($E_0 = 0.3\,\mathrm{eV}$, $r_0 = 35\,\mathrm{mm}$, $s = 0.3\,\mathrm{mm}$, $h_1 = 2\,\mathrm{mm}$, $h_2 = 6\,\mathrm{mm}$). The variation with the number of interactions is within the noise of calculation

Property	Number of iterations	Value
	1	3.66 nA
Optimum input current	2	3.84 nA
	3	3.75 nA
	1	1.09
Energy dispersion $E_0\delta\varrho/\delta E$	2	1.09
	3	1.09
	1	1.03
Magnification $|C_y|$	2	1.05
	3	1.02

Table 5.4. Noise test on the essential data of a monochromator as defined in Table 5.3, when the data are calculated repeatedly. The space charge potential is calculated with 500 trajectories randomly distributed in the realm of the parameters specifying the initial conditions. The transmitted energy distribution is calculated with 400 (random) trajectories at each incident energy, in energy intervals of 0.5 meV. Numbers in the last two rows give the mean and the standard deviation

I_{input} [nA]	I_{output} [nA]	ΔE_{FWHM} [meV]	$|C_y|$	$E_0\delta\varrho/\delta E$
3.65	0.455	2.87	1.02	1.10
3.75	0.451	2.70	1.00	1.14
3.75	0.466	2.75	1.04	1.09
3.83	0.464	2.80	1.00	1.09
3.63	0.444	2.66	1.05	1.08
3.89	0.472	2.69	1.00	1.09
3.75	0.459	2.74	1.02	1.10
±0.10	±0.10	±0.08	±0.02	±0.02

as for the deflector without a current load. This is already an important difference from the ideal 2D-cylindrical field. It is also useful to have information about the noise in the data due to the finite number of trajectories and the finite number of steps in their integration. Table 5.4 shows the essential results when the calculation is repeated.

We now present the results of the numerical analysis in more detail and begin with the trajectories in the limit of zero input current. The trajectories in the radial and vertical plane are displayed in Fig. 5.1a, b. The radial position of the electrons at the exit slit as a function of α_1 is shown in 5.1c for the three initial radial positions $r = r_0 - s_1/2$, $r = r_0$, and $r = r_0 + s_1/2$. The magnification $|C_y| \approx 1$ is clearly evident, as well as the typical quadratical dependence of y_2 on α_1. Because of the increased deflection angle, the focus is not at the exit slit (Fig. 5.1a). One therefore also has a linear term in $y_2(\alpha_1)$. In the vertical plane, electrons with $\beta_1 = 0$ display a small divergence although the potential on the top and bottom plates is nominally "zero", that is equal to the average of the potential of the inner and outer deflection plate, as are those of the entrance and

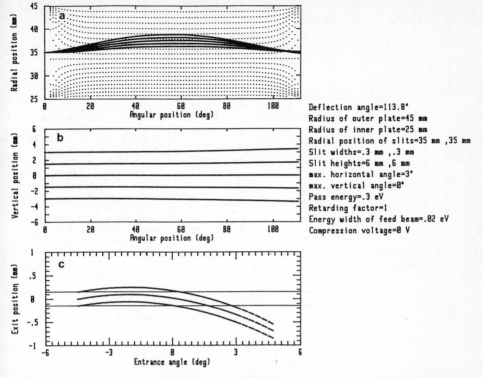

Deflection angle=113.8°
Radius of outer plate=45 mm
Radius of inner plate=25 mm
Radial position of slits=35 mm ,35 mm
Slit widths=.3 mm ,.3 mm
Slit heights=6 mm ,6 mm
max. horizontal angle=3°
max. vertical angle=0°
Pass energy=.3 eV
Retarding factor=1
Energy width of feed beam=.02 eV
Compression voltage=0 V

Fig. 5.1. (a) Trajectories in the radial plane of a cylindrical deflector, the deflection angle of which is extended by 7° ($\theta_f \approx 114°$) beyond the deflection angle leading to a first-order focus at the exit slit in the limit of zero current ($\theta_f \sim 107°$). Parameters of the deflector as in Table 5.1 (b) Trajectories in the vertical plane along the path of an electron entering the deflector with the pass energy at the radial coordinate r_0 and with entrance angles $\alpha = 0$, $\beta = 0$. The small divergence is caused by the fringe field of the upper and lower cover plates to which the average potential between inner and outer deflecting plate is applied. (c) Exit position of the electron as a function of the entrance angle for electrons entering the deflector at $r = r_0 + s_1/2$, $r = r_0$, and $r = r_0 - s_1/2$

exit slits. We note however from Fig. 5.1a that the electrons are incident at a potential that is negative with respect to the average potential. They therefore experience the potential of the bottom and top plates as a positive one and the small divergence of the beam shown in Fig. 5.1b results from this.

Figure 5.2 shows the trajectories of the electrons when the current load is such that the focus has shifted to the position of the exit, which is the case when the input current is 5.3 nA. The angular aberrations (Fig. 5.2c) now clearly have a third-order contribution. The angular aberration terms appear to have roughly the same values as those calculated for the 2D deflector (Table 4.2). The density of the potential grid in the 3D calculation is however not fine enough to permit a detailed analysis here. It is also evident from Fig. 5.2c that we have adjusted the input current in such a way that the difference between the exit positions of electrons with entrance angles of $-\alpha_{1m}$ and $+\alpha_{1m}$ is as small as possible, rather than adjusting the current so that the linear term in $y_2(\alpha_1)$ vanishes. As

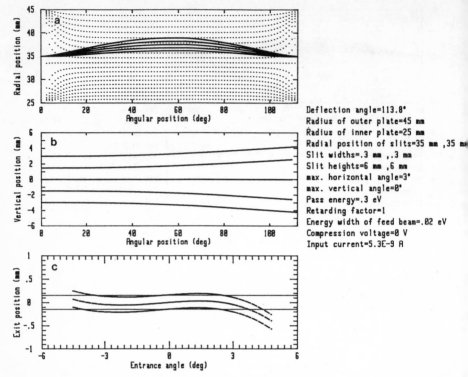

Deflection angle=113.8°
Radius of outer plate=45 mm
Radius of inner plate=25 mm
Radial position of slits=35 mm ,35 mm
Slit widths=.3 mm ,.3 mm
Slit heights=6 mm ,6 mm
max. horizontal angle=3°
max. vertical angle=0°
Pass energy=.3 eV
Retarding factor=1
Energy width of feed beam=.02 eV
Compression voltage=0 V
Input current=5.3E-9 A

Fig. 5.2. The same as Fig. 5.1, except that the current load now leads to a first-order focus at a deflection angle of $\theta_f \approx 114°$ (a). As a result of the space charge, the beam diverges along the z-direction (b). The exit position of the electron as a function of the entrance angle (c) displays a considerable third-order aberration although the total amount of the angular aberration is rather moderate and actually smaller than without space charge. The magnification remains $C_y \sim -1$, unlike the case of the ideal cylindrical field

discussed in Sect. 4.4, this procedure leads to a better resolution and transmission, at a somewhat higher input current also. As seen from Fig. 5.2b, the divergence of the beam in the θz-plane is now larger than in 5.1b, because of the repulsion by the space charge. When the exit and entrance slit are of the same height, this divergence results in a reduced transmission. The effect will become larger as the input current becomes larger, which will be the case for larger deflection angles. This is illustrated in Fig. 5.3 for extensions of the deflection angles up to $\Delta\theta_{sc} = 20°$. While the input current rises linearly as predicted by (4.73), the output current levels off and eventually saturates near 0.6 nA. It is also interesting to compare the numerical value for the input current with the theory. Using (4.73) and the parameters in Table 5.1, we find a value of 6.5 nA at $\Delta\theta_{sc} = 10°$. The current in Fig. 5.3 is higher. The reason for this small difference will be discussed later.

The divergence of the beam in the θz-plane and its adverse effect on the resolution may be compensated by compressing the beam with a small negative bias on the top and bottom plate. Figure 5.4 displays the trajectories in the

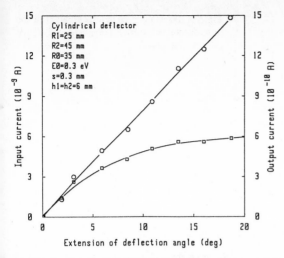

Fig. 5.3. Input and output current of a monochromator vs the extension of the deflection angle $\Delta\theta_{sc}$. The output current levels off because of the increasing divergence of the beam in the θz-plane

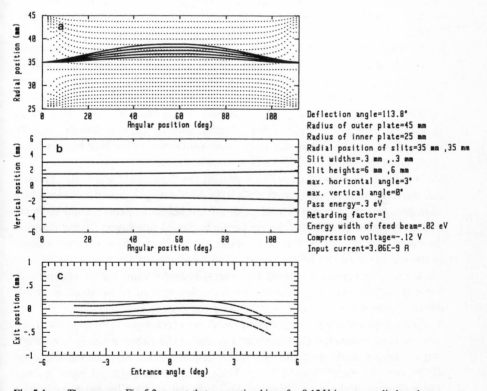

Deflection angle=113.8°
Radius of outer plate=45 mm
Radius of inner plate=25 mm
Radial position of slits=35 mm ,35 mm
Slit widths=.3 mm ,.3 mm
Slit heights=6 mm ,6 mm
max. horizontal angle=3°
max. vertical angle=0°
Pass energy=.3 eV
Retarding factor=1
Energy width of feed beam=.02 eV
Compression voltage=-.12 V
Input current=3.06E-9 A

Fig. 5.4a–c. The same as Fig. 5.2 except that a negative bias of $-0.12\,\mathrm{V}$ is now applied to the top and bottom cover plates. The divergence of the beam in the θz-plane due to the repulsive forces of the space charge is avoided

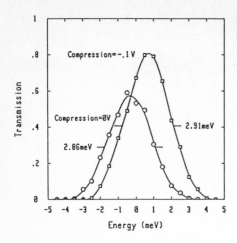

Fig. 5.5. Transmitted energy distribution of the monochromator (Table 5.1) for zero and −0.1 V compression bias

case when the compression voltage is −0.12 V. The trajectories in the θz-plane are now nearly straight lines. The optimum input current is reduced to 3.1 nA. This comes about because a compression voltage on the top and bottom plates shifts the focal angle θ to a larger value (Fig. 3.9). Thus less input current is needed to produce the radial focus at the exit slit. The compression obviously improves the transmission of the deflector. The latter is illustrated in Fig. 5.5 where the transmission is shown as a function of the energy of the electrons for compression voltages of zero and −0.1 V. The gain in transmission is about 40% and compensates the reduction of the input current. Since the resolution does not change with the compression (within the error margin) the effect of the compression voltage is mainly, (i) a shaping of the beam in the vertical plane, so that for example a better overall transmission in the spectrometer may be achieved, and (ii) an adjustment of the monochromator to the feed beam current provided at the entrance slit by a pre-monochromator.

Our results have so far been in line with the 2D model, which also predicts that the input and output currents scale with the slit height. Partly motivated by a consistent failure to find much effect of the slit height in experimental tests, we also investigated the effect of the slit height in the 3D simulation. To our surprise, we found that the diminution of the optimum input current is not proportional to the reduction of the height of the illuminated part of the entrance slit h_1 at all. Rather, the current levels off to a nearly constant value for h_1 smaller than 3 mm (Fig. 5.6). The output current has a shallow maximum for small h_1, while the resolution remains nearly constant as a function of h_1.

Since the simulation was performed with zero compression voltage, the transmission decreases when the illuminated height of the entrance slit h_1 is enlarged and the height of the exit slit h_2 remains at 6 mm. This explains the reduction of the output current beyond $h_1 > 3$ mm, despite the enlarged input current. The reduction of the output current can hardly be avoided by again applying a compression voltage as in Figs. 5.4 and 5.5 because the optimum input current then becomes smaller.

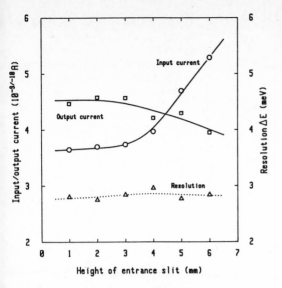

Fig. 5.6. Input current, output current, and resolution (FWHM) for the monochromator as a function of the illuminated height of the entrance slit h_1

In interpreting the results displayed in Fig. 5.6, we are left with the question: why is the optimum input current not proportional to h_1, as predicted by the 2D model? The basic assumption of the 2D model is that the height of the beam along the z-axis is large compared with its extension in the radial plane. This is obviously true in the vicinity of the entrance slit. Within the deflector, however, two-dimensionality also requires that (4.20)

$$h_1 \gg \frac{r_0 \alpha_{1m}}{\sqrt{2}} \sin \sqrt{2}\theta + r_0 \frac{\Delta E_{in}}{2E_0}(1 - \cos \sqrt{2}\theta) \,. \tag{5.7}$$

Taking the maximum values of each term for the parameters listed in Table 5.1 it would be required that

$$h_1 \gg 1.29\,\text{mm} + 2.33\,\text{mm} \,. \tag{5.8}$$

Clearly this condition is not fulfilled with $h_1 \leq 6\,\text{mm}$. In fact, a levelling off of the input current to a constant value for h_1 smaller than $\sim 3\,\text{mm}$ is not so surprising when one considers (5.8). For small h_1, one eventually approaches the limit where the space charge is confined to a sheet and the repulsive forces on electrons travelling on boundary trajectories in such a sheet are independent of the height of the sheet. Figure 5.6 also shows us that "two-dimensional" behaviour cannot be achieved in realistic designs. Apart from manufacturing problems with higher aspect ratios, one would also need to have a lens system between the monochromator and the analyser capable of forming a good image of an extended slit. Such lenses are however not available, with electrostatic lenses at least. Since the value of h_1 at which the current becomes nearly independent of h_1 depends on α_{1m} and ΔE_{in} and since these parameters are also fixed by other considerations, it follows that the illuminated slit height h_1 should be about 2 mm

73

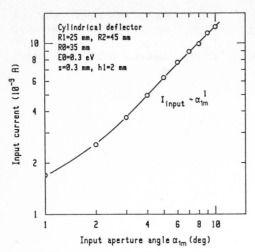

Fig. 5.7. Optimum input current vs the aperture angle of the feed beam for the cylindrical deflector defined in Table 5.1. The linear relation is in accordance with (4.73) derived for a two-dimensional model of the space charge

within our chosen reference frame of parameters. The precise value is of course not very critical. Smaller values of h_1 are also possible, when the monochromator alone is considered. We shall come back to this issue in the context of electron emission systems.

The considerable departure of the three-dimensional deflector from the two-dimensional model makes it necessary to explore the dependence of its properties on the other parameters as well. A comparison of the equation for the optimum input current in the two-dimensional case (4.73) and the analytical condition for the slit height h (5.7) required to make (4.73) applicable seems to suggest that the dependence of the input current on the radius r_0 in (4.73) also vanishes when the current becomes independent of the slit height h. In fact, the general scaling of space-charge-saturated currents requires that the total current should not depend on the absolute size of the device, (2.20). The remaining parameters one needs to study are hence the aperture angle α_{1m} and the energy width ΔE_{in} of the feed beam. Figure 5.7 shows the dependence of the optimum input current on the aperture angle α_{1m}. Except for very small α_{1m}, the input current displays the linear behaviour as predicted by the two-dimensional model and (4.73). The deviation at small α_{1m} obviously comes about because one leaves the realm of α-values in which the condition $\alpha_{1m}/\sqrt{2} > \Delta E_{in}/2E_0$ holds.

The output current is shown as a function of the aperture angle α_{1m} in Fig. 5.8. The output current deviates from the linear relation because of the decreasing transmission for large α_{1m} (compare Sect. 3.3). Simultaneously, the width of the energy distribution of the transmitted electrons increases. Since the output current generally scales with the transmitted energy width as $\Delta E_{1/2}^{5/2}$, one may define a performance factor

$$c_{p_{out}} = \frac{I_{out}}{\Delta E_{1/2}^{5/2}} . \tag{5.9}$$

This performance factor is also plotted in Fig. 5.8. In order to reduce the noise

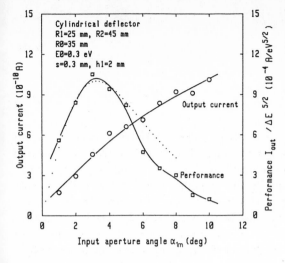

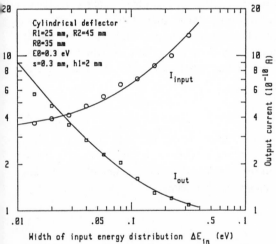

Fig. 5.8. Output current and performance factor $I_{out}/\Delta E_{1/2}^{5/2}$ vs the angular aperture α_{1m} of the feed beam for the cylindrical deflector defined in Table 5.1. The maximum of the performance factor occurs near 3.5°, approximately where predicted for the ideal cylindrical field without space charge. The dotted line indicates the calculated performance factor (in arbitrary units) for the ideal cylindrical field

Fig. 5.9. Optimum input and output current vs the energy width of the feed beam for the cylindrical deflector defined in Table 5.1. The base width of the transmitted energy distribution remains about constant. The solid line for the input current corresponds to a fit with $I_{input} = 0.6 \; k \; E_0^{3/2} \Delta\theta_{sc}(\alpha_{1m} + \alpha_c)$ where $\alpha_c = 0.16 \; \Delta E_{in}/E_0$. The solid line for the output current represents the equation $I_{out} = 0.42 \; I_{input} \Delta E_{B\,out}/\Delta E_{in}$

in the data points, we have taken $\Delta E_{1/2}$ to be half the base width ΔE_B. The latter quantity is more easily calculated with high precision. The performance factor has a relatively sharp maximum near $\alpha_{1m} \sim 3.5°$, which is consistent with (3.38) derived for the cylindrical deflector without space charge. The shape of the curve of the performance factor is also quite similar to the curve calculated for the cylindrical deflector without space charge following (3.33) and (3.35) (dotted line in Fig. 5.8).

For a full exploration of the relevant parameters of a space-charge-optimised deflector, we need to establish the dependence of the performance on the energy width of the feed beam. We have again calculated the results for our standard deflector specified in Table 5.1 and the results are displayed in Fig. 5.9. The rise of the optimum input current with the width of the energy distribution in the feeding current is according to expectation. The input current may be described by the equation

$$I_{input} = 0.6 \, k \, E_0^{3/2} \Delta\theta_{sc}(\alpha_{1m} + \alpha_c) \tag{5.10}$$

with the critical angle

$$\alpha_c = 0.16 \frac{\Delta E_{in}}{E_0} . \tag{5.11}$$

The solid line in Fig. 5.9 represents (5.10) with (5.11). The two equations also fit the dependence on α_{1m} as displayed in Fig. 5.7. Comparison with (4.74) shows that when the space charge is confined to a sheet of small height as studied here, the factor $4h/r_0$ appearing in the two-dimensional model (4.74) is now replaced by the prefactor 0.6 in (5.10). We again recover the result that the transition to the region in which the behaviour is that predicted by the two-dimensional model lies above $h \approx 5$. Figure 5.9 also shows the output current vs ΔE_{in}. The prefactor drops further to 0.35 when a compression voltage is applied, such that the trajectories remain parallel in the vertical plane under space charge conditions. The required compression potential is about 0.55 E_0 for the layout of the monochromators used here. This compression potential causes the parallel beam at the entrance slit to form a focus close to the exit slit when the current is zero. Provided that the extension of the pass length $\Delta\theta$ is not too large, one may describe the output current by

$$I_{out} = T_\alpha T_E I_{input} , \tag{5.12}$$

where T_α and T_E are the transmission factors as defined in (3.30) and (3.48).

In T_α one may insert the aperture angle where one has the maximum performance (Fig. 5.8). We have seen that the $\alpha_{1m\,opt}$ is approximately as calculated for the ideal field without space charge (3.39), namely

$$\alpha_{1m\,opt} = \sqrt{\frac{s}{2r_0}} . \tag{5.13}$$

Taking this together with the approximate equation for the FWHM of the transmitted energy distribution (3.56),

$$\frac{\Delta E_{1/2}}{E_0} = \frac{s}{r_0} + 0.47 \, \alpha_{1m}^2 , \tag{5.14}$$

we finally obtain an expression for the monochromatic current as a function of the resolution

$$I_{out} \approx 0.14 \, k \frac{r_0}{s} \frac{\Delta E_{1/2}^{5/2}}{\Delta E_{in}} \Delta\theta_{sc} \left(1 + 0.28 \sqrt{\frac{s}{r_0}} \frac{\Delta E_{in}}{\Delta E_{1/2}} \right) . \tag{5.15}$$

The desired high monochromatic current obviously calls for a large ratio r_0/s The finite space in vacuum chambers places an upper limit on the radius r_0. In making the slit width smaller one eventually encounters the limit that the cathode

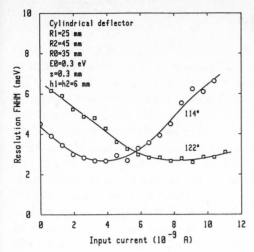

Fig. 5.10. Resolution of monochromators with deflecting angles of 114° and 122° vs input current

cannot provide the amount of input current into the entrance slit that could be handled by the monochromator. The remaining tunable design parameter is the extension of the deflection angle $\Delta\theta_{sc}$.

A prudent choice of the extension of the deflection angle is of eminent importance for the performance. In the course of our numerical exploration of many different spectrometers, we have enquired whether the calculations suggest a definite optimum value of $\Delta\theta_{sc}$. Such an optimum would, for example, exist very clearly if the energy resolution became poorer at higher deflection angles. It appears however, that the increased energy dispersion with larger deflection angles balances the higher aberrations (Table 4.2), which are also associated with larger deflection angles. Thus no natural limit for the extension of the deflection angle arises from these calculations alone. Some guidance for the choice is provided by Fig. 5.10 where we have plotted the FWHM of the transmitted energy distribution vs the input current for two different deflection angles, 114° and 122°. Both curves display a relatively shallow minimum in the FWHM, with the 122° deflector providing the higher current. The deflection angle of 122° is superior, however, only if one has an input current available of about 8 nA (at 20 meV energy width). As we shall see in the next section, this current approaches the theoretical limit for the output current of a pre-monochromator. Because of the power law of the current at the detector, resolution is also precious. One should therefore choose a deflection angle for a monochromator for which the loss in resolution for smaller feed currents remains reasonable. With this in mind, a deflecting angle of about 115° for the second monochromator is proposed as a good choice when the ratio of the radii of the outer and inner deflecting plate R_2/R_1 is 1.8 as assumed here.

5.2 Retarding Monochromators

In this section we present results of the numerical simulation of monochromators with space charge when either the exit aperture is at a reduced potential or the entrance slit at a higher potential. As in the previous section, we first list the design parameters (Table 5.5). The deflection angle is extended by a larger amount now, since for a pre-monochromator a possible loss in resolution due to an imperfect match of the input current does not affect the resolution of the second monochromator. A variable resolution is therefore tolerable. The mean radius r_0 is reduced to the smaller value of 25 mm, mainly in the interest of a compact design (see however also Chap. 8). The radial position of the exit slit is slightly shifted outwards in order to keep the angles α_2 of the exiting electrons centred around $\alpha_2 = 0$ (Sect. 3.2). The width of the entrance slit is reduced to match the larger magnification. The aperture angle α_{1m} is also reduced by the same factor (Sects. 3.18 and 3.20). For retarding deflectors we define the retardation factor as

$$F_{\text{exit}} = \frac{E_0}{eU_{\text{exit}}} \qquad (5.16)$$

with eU_{exit} the potential energy of the electrons at the exit slit and E_0 the nominal pass energy. We have chosen the retardation factor to be 5 in keeping with the optimisation of the match of the resolution of the pre-monochromator to that of the second monochromator (Table 5.2). With a nominal pass energy of 1.5 eV, the potential energy at the exit aperture is therefore 0.3 eV. The nominal pass energy E_0 is now defined by the difference in the potentials of inner and outer deflection plates according to

$$eU(R_2) - eU(R_1) = 2E_0 \ln(R_2/R_1) . \qquad (5.17)$$

Table 5.5. List of design parameters for a reference, decelerating monochromator

Deflection angle	θ_{sc}	127°
Radius of outer deflection plate	R_2	31.7 mm
Radius of inner deflection plate	R_1	17.6 mm
Total height	H	44 mm
Radial position of entrance slit	r_{01}	25 mm
Radial position of exit slit	r_{02}	25.5 mm
Width of entrance slit	s_1	0.15 mm
Width of exit slit	s_2	0.3 mm
Height of entrance slit	h_1	1–6 mm
Height of exit slit	h_2	6 mm
Maximum aperture angle α	α_{1m}	1.5°
Maximum aperture angle β	β_{1m}	0°
Energy width of feed beam	ΔE_{in}	0.2 eV
Pass energy	E_0	1.5 eV
Retardation factor	F	5
Potential on top and bottom plates	V_0	0.3 V

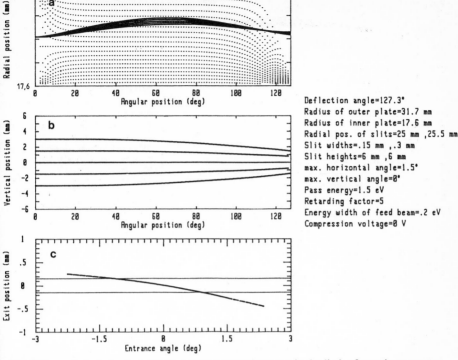

Deflection angle=127.3°
Radius of outer plate=31.7 mm
Radius of inner plate=17.6 mm
Radial pos. of slits=25 mm ,25.5 mm
Slit widths=.15 mm ,.3 mm
Slit heights=6 mm ,6 mm
max. horizontal angle=1.5°
max. vertical angle=0°
Pass energy=1.5 eV
Retarding factor=5
Energy width of feed beam=.2 eV
Compression voltage=0 V

Fig. 5.11a–c. Trajectories in a 127.3° retarding monochromator in the limit of zero input current. The retardation factor is $F_{exit} = 5$, see equipotential lines dotted in (**a**). The focus in the radial plane is near 111.5°. The trajectories in the θz-plane (**b**) are convergent because the potential of the top and bottom plates is equal to the potential of the exit slit and thus negative with respect to the pass energy. (**c**) The exit position y_2 as a function of α_1. Evidently the linear term in $y_2(\alpha_1)$ does not vanish

E_0 would be the real pass energy if the field were that of the ideal cylindrical field (3.1). For the monochromator with a retardation factor of $F = 5$ and the exit slit moved outwards, as described above, we find the pass energy to be higher than E_0 by a factor of ~ 1.12. For a nonretarding monochromator with entrance and exit slits of the same radius, the difference between the nominal pass energy as defined by (5.17) and the actual pass energy is merely about 1%. That difference was therefore disregarded in the previous considerations. In the retarding monochromator to be discussed here we have set the potential on the top and bottom plates equal to that of the exit aperture, which means that they are negatively biased with respect to the potential in the centre of the device.

As with the nonretarding monochromator, we first show the trajectories in the limit of zero input current (Fig. 5.11). The radial focus is near 111.5° in that case so that the actual extension of the deflection angle is $\sim 16°$. In the θz-plane the beam is compressed (Fig. 5.11b) because of the potential on the top and bottom plates. When the deflector is fed with a current chosen to shift the radial focus into the exit slit (Fig. 5.12a), the beam is less compressed in the θz-plane

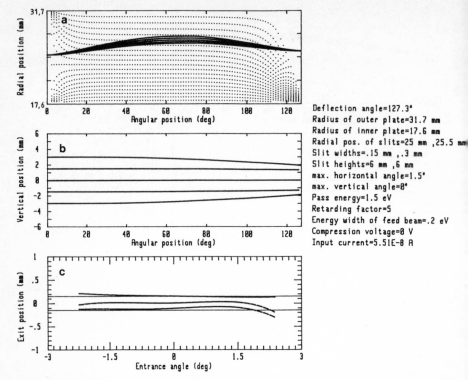

Deflection angle=127.3°
Radius of outer plate=31.7 mm
Radius of inner plate=17.6 mm
Radial pos. of slits=25 mm ,25.5 mm
Slit widths=.15 mm ,.3 mm
Slit heights=6 mm ,6 mm
max. horizontal angle=1.5°
max. vertical angle=0°
Pass energy=1.5 eV
Retarding factor=5
Energy width of feed beam=.2 eV
Compression voltage=0 V
Input current=5.51E-8 A

Fig. 5.12. The same retarding monochromator as in Fig. 5.8, but now the input current shifts the radial focus into the exit slit (a). The trajectories in the θz-plane are less convergent here (b) because of the repulsive potential caused by the space charge. The exit position of the electrons as a function of α_1 (c) shows a remarkably low angular aberration. The three curves in (c) correspond to initial radial positions of $r = r_0 - s_1/2$, $r = r_0$ and $r = r_0 + s_1/2$. Note however that s_1 is 0.15 mm while s_2 is 0.3 mm. The magnification is $C_y = -1.75$

(Fig. 5.12b). The optimum potential on the top and bottom plates, where the trajectories become nearly straight lines in the θz-plane, is somewhere between the potential of the exit aperture and the potential corresponding to the nominal pass energy. The angular aberrations of the retarding monochromator subject to space charge are quite small (Fig. 5.12c). The magnification C_y is -1.75, while it would be about -1.5, if the deflector were not extended (3.18). We have again investigated input current, output current and resolution as a function of the illuminated height of the entrance slit h_1 (Fig. 5.13) and find a similar result to that for the nonretarding monochromator. The output current changes relatively little with h_1. Once again, therefore, it seems preferable to use a slit height of about 2 mm. For a slit height of $h_1 = 2$ mm, Fig. 5.14 displays the trajectories. They are practically straight lines in the θz-plane when the top and bottom plates are held at the potential of the exit slit.

Hitherto we have investigated retarding monochromators with the exit slit at a lower potential than that corresponding to the nominal pass energy E_0, while

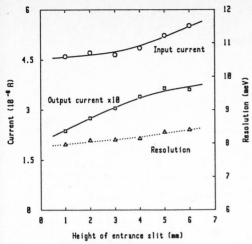

Fig. 5.13. Optimum input current, output current and the FWHM of the transmitted energy distribution for the 127° retarding monochromator as a function of the illuminated height of the entrance slit. Just as for the nonretarding monochromator, the gain in output current with larger entrance slit is marginal

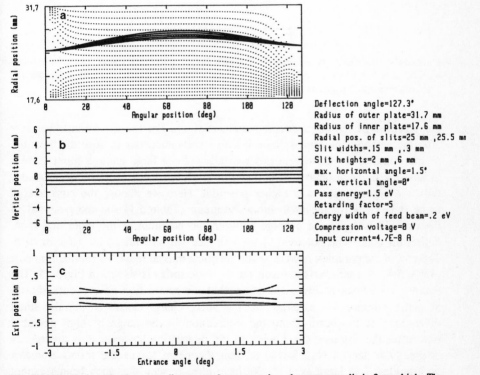

```
Deflection angle=127.3°
Radius of outer plate=31.7 mm
Radius of inner plate=17.6 mm
Radial pos. of slits=25 mm ,25.5 mm
Slit widths=.15 mm ,.3 mm
Slit heights=2 mm ,6 mm
max. horizontal angle=1.5°
max. vertical angle=0°
Pass energy=1.5 eV
Retarding factor=5
Energy width of feed beam=.2 eV
Compression voltage=0 V
Input current=4.7E-8 A
```

Fig. 5.14a–c. Trajectories in a retarding monochromator when the entrance slit is 2 mm high. The trajectories in the θz-plane (b) run nearly parallel when the top and bottom plates have the same potential as the exit slit

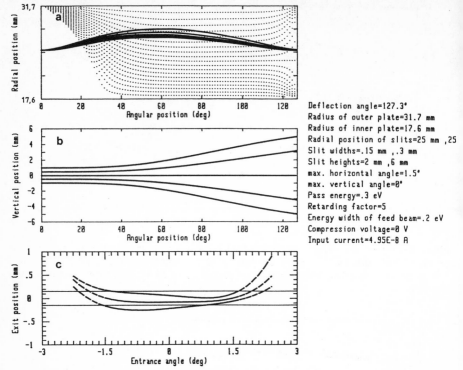

Deflection angle=127.3°
Radius of outer plate=31.7 mm
Radius of inner plate=17.6 mm
Radial position of slits=25 mm ,25
Slit widths=.15 mm ,.3 mm
Slit heights=2 mm ,6 mm
max. horizontal angle=1.5°
max. vertical angle=0°
Pass energy=.3 eV
Retarding factor=5
Energy width of feed beam=.2 eV
Compression voltage=0 V
Input current=4.95E-8 A

Fig. 5.15a–c. Trajectories in a cylindrical deflector loaded with space charge when the entrance aperture is at five times the pass energy. Since the retardation occurs shortly after the entrance aperture the beam diverges extensively in the θz-plane (**b**)

the potential of the entrance slit remained at E_0. For the first monochromator after the cathode emission system, it may be advantageous to raise the potential of the entrance slit further in order to allow for a large enough input current to feed the monochromator. It is therefore useful to study a pre-monochromator with the entrance slit at a higher potential. Here we choose the pass energy to be 0.3 eV, as for the main monochromator (Table 5.1) and the potential of the exit slit is again the average between the potential of the inner and outer deflection plate. The potential at the entrance slit was raised by a factor of 5. The rest of the parameters were as for the conventional retarding monochromator (Table 5.4). A characteristic result for the trajectories is shown in Fig. 5.15. The equipotential lines in Fig. 5.15a show that the retardation now occurs shortly after the entrance slit, as expected. The space charge causes an initially small divergence of the beam along the z-direction in the range of high potential right after the entrance slit. Owing to the retardation however, this divergence is greatly enhanced (Fig. 5.15b) so that the beam leaves the monochromator with angles β as large as $\pm 5°$. We have already seen that such beams cannot be effectively processed further in a second monochromator or a lens system. Monochromators with higher potentials at the entrance slit therefore require a

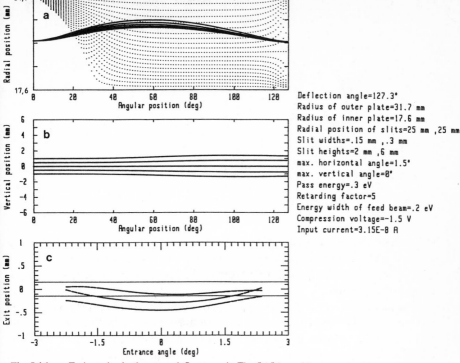

Deflection angle=127.3°
Radius of outer plate=31.7 mm
Radius of inner plate=17.6 mm
Radial position of slits=25 mm ,25 mm
Slit widths=.15 mm ,.3 mm
Slit heights=2 mm ,6 mm
max. horizontal angle=1.5°
max. vertical angle=0°
Pass energy=.3 eV
Retarding factor=5
Energy width of feed beam=.2 eV
Compression voltage=-1.5 V
Input current=3.15E-8 A

Fig. 5.16a–c. Trajectories in the same deflector as in Fig. 5.15 but with a negative bias applied to top and bottom cover plates. The optimum input current is reduced to $\sim 30\,\text{nA}$

relatively large negative bias on the top and bottom cover plates in order to keep the beam from diverging along the z-axis. Figure 5.16 shows the trajectories in the same deflector as in Fig. 5.15 but with a compression voltage of $-1.5\,\text{V}$ applied to top and bottom cover plates. The transmission is now high again, angular aberrations appear to be low, and the trajectories leave the deflector at positions and angles suitable for further processing. Compared with the analyser with zero compression voltage (Fig. 5.15), the optimum input current is reduced to 31 nA instead of 49 nA. The output current is however about the same, thanks to the higher transmission. The transmitted energy distribution is well shaped (Fig. 5.17) and provides for a respectable 3.1 meV resolution at a current of $\sim 0.4\,\text{nA}$. We therefore realise that retarding monochromators can be used in both ways, either with the entrance aperture at a higher or the exit aperture at a lower potential. When intended as pre-monochromators, one should allow the deflector to be operated in either way, which requires no extra lead into the vacuum system, but merely one extra adjustable potential.

Pre-monochromators in our instruments typically supplied higher currents when the potential at the entrance aperture was raised to about double the pass energy of the device. Presumably the main effect of the higher potential at the entrance slit is to provide higher target potential for the cathode emission system

83

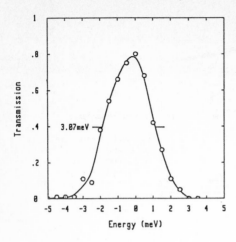

Fig. 5.17. Transmitted energy distribution of the cylindrical deflector specified in Fig. 5.16. The output current is $\sim 0.4\,nA$

and thereby reduce the space charge in the cathode emission system. Returning to (1.19), we may estimate that passing currents above 10 nA into a slit of 0.15×2 mm at an energy of 1.5 eV takes us well into the range of space-charge-limited feed currents.

The numerical results for the current and resolution are in keeping with (5.15) when the width of the entrance slit is substituted for s in (5.15). Note that the entrance slit is reduced in size in order to match the magnification. We likewise find the current of the retarding deflector with exit slit retardation to be well described by (5.15). The set of equations (5.10–15) may therefore be used for an optimum match of the pre-monochromator to the second monochromator. Depending on the slit widths, the radii, and the extension of the deflection angle the result differs slightly from the result obtained with the simplified procedure following (5.5) and Table 5.2. The difference is however of no practical consequence because of the considerable degree of flexibility one builds into a spectrometer by making enough potentials tunable.

Such a fine tuning could, for example, be applied to the second monochromator specified in Table 5.1. One may again have the potential at the exit or the entrance slit variable. When the potential of the exit slit is lowered the optimum input current is reduced (Fig. 5.18). At the same time the width of the transmitted energy distribution also shrinks. The performance factor as defined by (5.1) remains constant however. A different result is obtained when the potential of the entrance slit is raised (Fig. 5.19) with respect to the pass energy. The optimum input current then increases and the width of the transmitted energy distribution also becomes slightly larger. Nevertheless the performance factor with respect to the entrance current rises with higher potentials at the entrance slit. When only a single monochromator is used in a spectrometer it may therefore be advantageous to have the entrance slit at a higher potential. The price to be paid is that the beam diverges more in the θz-plane. When a negative bias to the cover plates is applied simultaneously in order to keep the beam confined in the θz-plane, the performance becomes independent of the retardation factor. When used as a

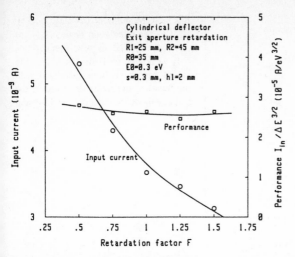

Fig. 5.18. Optimum current and performance of a retarding deflector as specified in Table 5.1 as a function of the retardation factor F. The lower the potential at the exit aperture ($eU_{\text{exit}} = E_0/F_{\text{exit}}$) the lower the optimum input current. The resolution however increases also. The performance constant $c_p = I_{\text{input}}/\Delta E^{3/2}$ thus remains constant

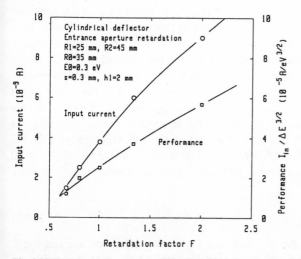

Fig. 5.19. Optimum input current and performance of a retarding deflector when the entrance aperture is at higher or lower potential. The input current rises with the potential at the entrance aperture $eU_{\text{entr}} = E_0 F_{\text{entr}}$, the resolution becomes a little lower. The performance constant c_p increases however with F_{entr}

second monochromator, adjustment of the potential at the entrance aperture of the second monochromator (which is also the exit aperture of the first) serves to provide an optimum match between the output current of the first monochromator and the most favourable input of the second. We remember that this match

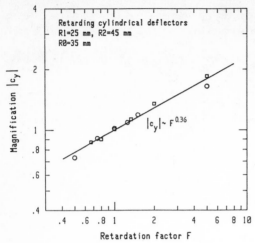

Fig. 5.20. Modulus of the magnification $|C_y|$ vs the retardation factors F_{exit} (o) and F_{entr} (□). The magnification depends only weakly on the deflection angle. The values here have been calculated with the retarding deflectors described in the text having deflection angles of 114° and 127°

can also be achieved by adjusting the ratio of the pass energies of the first and second monochromator. It is therefore not really necessary to have the aperture of the second monochromator at a variable potential.

We conclude this section with a plot of the magnification C_y as a function of the retardation factor when the monochromator is loaded with the optimum input current. When designing a spectrometer, with or without pre-monochromator, the layout should roughly fix the retardation factors in the pre-monochromator or – if one settles for a single stage monochromator - in the main monochromator. Since the retardation factors also affect the magnification, the ratio of the width of the entrance and exit slits should be matched appropriately. In Fig. 5.20 the modulus of the magnification is plotted vs the retardation factor. The result is within the numerical error the same for deflectors with entrance or exit slit retardation, and also for the different deflection angles considered. Figure 5.20 shows that the magnification for the deflectors loaded with space charge roughly scales according to

$$|C_y| = F^{0.35} , \tag{5.18}$$

a result which is close to the one obtained earlier for deflectors without space charge (3.18). We have already mentioned in Sect. 3.2 the fact that, with the magnification, the angular aperture also changes (3.19) by an amount that can be calculated by applying the rule of phase space conservation in two dimensions.

6. Electron Emission Systems

Monochromators require feed beams with defined aperture angles and a low energy spread. After presenting some basic concepts of thermionic emission we describe the technical aspects of trajectory calculations in the presence of space charge. Emission systems with differently shaped lenses are studied and their performances are compared.

6.1 Basic Concepts

In the preceding chapter we have investigated the maximum monochromatic current provided by cylindrical deflectors. The result of these considerations could be expressed by (5.15), describing the monochromatic current of both nonretarding and retarding monochromators. The geometric parameters of the deflector that appear there are the extension of the deflection angle and the ratio of the centre radius and the slit width. In order to produce the monochromatic currents predicted by (5.15) and the associated numerical analysis, the monochromator needs to be fed with a beam of electrons produced by an emission system whose beam parameters must meet the specifications required by the monochromator. We state these specifications here more explicitly by combining (5.12) and (5.15), whereupon we obtain an expression for the input current of an monochromator:

$$I_{input} \approx 0.18 k \frac{r_0}{s} \Delta E_{1/2}^{3/2} \Delta\theta_{sc} \left(1 + 0.28 \sqrt{\frac{s}{r_0}} \frac{\Delta E_{in}}{\Delta E_{1/2}}\right) . \tag{6.1}$$

This expression tells us that a monochromator can accept a higher feed current and thus provide higher monochromatic currents the larger the ratio r_0/s is. For a given overall size of the device, the monochromatic current could be improved by reducing the slit width, which would in turn require a higher current density of the feed beam. As a subsidiary condition to (6.1), we have shown in Sect. 5.1 that the angular aperture α_{1m} should be close to or smaller than

$$\alpha_{1m} = \sqrt{\frac{s}{2r_0}} . \tag{6.2}$$

A similar subsidiary condition governs the angular aperture β_{1m}, which we take from (3.39) with $n = 5/2$

$$\beta_{1m} < \sqrt{\frac{2s}{3r_0}} .$$ (6.3)

While this condition arose from the optimisation of the transmission with respect to resolution, a small angular divergence is also important with regard to the space charge because a small β_{1m} allows us to compensate the space charge spreading by applying a negative bias to the bottom and top deflecting plates. A small angle β_{1m} is also important for the lens system beyond the monochromator, more important than a small divergence in α as we shall see later. Finally the illuminated slit height is subject to a further constraint when the quantities r_0, α_{1m}, and ΔE_{in} are the scaling parameters (Sect. 5.7). Because of these side conditions, scaling up to the ratio r_0/s in order to achieve higher monochromatic currents entails focusing the feed beam into a smaller area *and* into a smaller angular aperture. Furthermore, it is obvious that the feed beam should have as narrow an energy distribution as possible. With these considerations, we have therefore established the current per unit area, solid angle, and energy window at the entrance slit as an important quantity for the evaluation of an electron emission system. The quantity is usually referred to as the "brightness". We define the brightness B here as

$$B = \frac{d^2 j}{d\Omega \, dE} ,$$ (6.4)

where j denotes the current density. In the older literature (see e.g. [6.1]), the term brightness is used for the current per solid angle and area. The definition of brightness adopted here, which also includes the energy spread, is in line with modern optical practice, when the properties of a synchrotron light source are described for example [6.2]. To give an example, the brightness needed to feed the pre-monochromator specified in Table 5.5 is of the order of 1 A/(eV cm^2 rad) if we assume β_{1m} to be 1°. In order to relate this quantity to the established properties of electron sources, we need to consider some basic concepts of electron emission. We restrict ourselves to thermionic emission. Field emission sources, while providing a higher brightness, have not been tamed to feed high resolution monochromators so far. The following material is to be found in the standard books on electron optics [6.1, 3] and is repeated here in our notation for easier reference.

Firstly, we are interested in the energy distribution of emitted electrons, which is the Maxwellian flux distribution. The differential current density into a velocity window dv and into the solid angle $d\Omega = \sin \vartheta \, d\vartheta \, d\varphi$ is

$$d^2 j = 2e \left(\frac{m}{h}\right)^3 v^3 \exp \left(-\frac{mv^2 + e\Phi}{2k_B T}\right) \cos \vartheta \, d\Omega \, dv ,$$ (6.5)

where $e\Phi$ denotes the work function of the emitter and k_B the Boltzmann constant. The prefactor disregards reflection of the electrons from the surface barrier and therefore represents an upper limit. After converting this into an energy distribution, one obtains for the brightness

$$\frac{d^2 j}{d\Omega \, dE} = 4e \left(\frac{m}{h}\right)^3 E \exp\left(-\frac{E + e\Phi}{k_B T}\right) \cos\vartheta \ . \tag{6.6}$$

We note that this is the energy distribution of the flux, not the energy distribution of the density, which would be proportional to $\sqrt{E} \exp(-E/k_B T)$. The relevant energy dependence of the brightness $E \exp(-E/k_B T)$ has its maximum at

$$E_m = k_B T \cong T \frac{eV}{11600\,K} \ . \tag{6.7}$$

The highest brightness of a beam in an aperture is obviously obtained when the lenses focus electrons at the maximum of the energy distribution. The fraction of the current falling into a narrow energy window dE relative to the total current emitted into the same solid angle is also of interest

$$\frac{dj}{j\,dE} = \frac{E_m \exp\left(-E_m/k_B T\right)}{\int_0^\infty E \exp\left(-E/k_B T\right) dE} \ , \tag{6.8}$$

$$\frac{dj}{j\,dE} = \frac{e^{-1}}{k_B T} \ , \qquad (e = 2.718) \ . \tag{6.9}$$

This equation allows us to calculate the current in a small energy window of a few meV when a monochromator is fed with a particular current load stemming from thermionic emission. The current per energy interval falling into the angular apertures accepted by a monochromator imposes an upper bound on the monochromatic current that a monochromator can deliver. Taking our standard monochromators as examples, currents per energy interval of 0.2–0.4 nA/meV are needed (Figs. 5.6, 13). We finally recall that the full width at half maximum of the Maxwellian flux distribution is given by

$$\Delta E_{FWHM} \approx 2.45 k_B T = T \frac{eV}{4740\,K} \ . \tag{6.10}$$

The brightness of an electron source and the brightness at any aperture further along the electron optical system are related because of phase space conservation. For beams with small angular apertures phase space conservation means that the product of energy, solid angle, and cross-section of the beam remains constant. Applying this principle to electron emission from the cathode and the image of the cathode surface area we obtain

$$F_{cathode} \, d\Omega_{cathode} \, E_{cathode} = F_{image} \, d\Omega_{image} \, E_{image} \ . \tag{6.11}$$

The energy at the cathode may be replaced by the energy where the Maxwellian distribution has its maximum i.e., where the source has the highest brightness. We then obtain

$$F_{cathode} \, d\Omega_{cathode} \, k_B T = F_{image} \, d\Omega_{image} \, (eV_{ac} + k_B T) \ , \tag{6.12}$$

where V_{ac} is the acceleration voltage. In addition to volume in phase space, current and energy are also conserved. The width of the energy distribution in

89

the beam therefore remains the same. (We disregard the Boersch effect, which is important for large acceleration voltages and beams of high current density [6.4]). Because of the conservation of current and the conservation of the energy distribution, one has conservation for the product of the brightness B, area and the solid angle

$$B_{cathode} \, F_{cathode} \, d\Omega_{cathode} = B_{image} \, F_{image} \, d\Omega_{image} \qquad (6.13)$$

and therefore

$$B_{image} = B_{cathode} \frac{(eV_{ac} + k_B T)}{k_B T} \,. \qquad (6.14)$$

This conservation law for the brightness is familiar from normal light optics. There, the ratio of the energies is replaced by the ratio of the refractive indices of the media in which the light travels at the source and image position. In order to compare the brightness of a typical tungsten emitter with the brightness of about $1 \, A/eV \, cm^2$ rad required to feed a monochromator we list the brightness of such an emitter in Table 6.1 for various cathode temperatures. Since the brightness in the entrance slit of the monochromator is enhanced by the factor $(eV_{ac}+k_B T)/k_B T$ relative to the brightness of the cathode, it seems that a tungsten cathode could be operated at temperatures below 2400 K. It also appears that moving to sources of higher brightness would not have any advantage. In these considerations we have however disregarded the properties of the lens system and space charge effects. We shall therefore need to revisit the issue of source brightness after discussing some concrete emission systems.

Table 6.1. Brightness of a tungsten emitter (After [6.3])

T[K]	$B \, [A/eV \, cm^2 \, rad]$
2000	2.1×10^{-3}
2200	7.0×10^{-2}
2400	0.56
2600	3.2
2800	14.6

6.2 Technical Aspects of the Calculations

In this section we present a few technical aspects of the calculations of lens systems, including the modification of the trajectories by the space charge and the chromatic error. The details may be particularly useful for readers who intend to perform similar calculations for themselves.

The numerical calculations are delt with in two separate programs. The first defines the lens elements and solves the Laplace equation for arbitrary potentials on the lens elements, using the superposition principle as described in Sect. 2.2.

The second program calculates the trajectories, the space charge potential, and all the other electron optical properties of interest.

The solution of the Laplace equation was performed with a mesh size of $0.5 \times 0.5 \times 0.5 \, \text{mm}^3$ in cartesian coordinates. Because of the symmetry of the element, only one quadrant need be calculated. The typical size of the array was then $30 \times 40 \times 40$. We have also experimented with a finer mesh of $0.25 \times 0.25 \times 0.25 \, \text{mm}^3$. The differences in the resulting trajectories were so small that we reverted to the larger mesh size for all but a few trial calculations in the interest of shorter computing times. When the Laplace algorithm is applied to one quadrant of the lens, the procedure must be modified along the dividing planes between two quadrants and also along the optic axis common to all four quadrants. We illustrate the procedure with the yz-plane when z is the optic axis. We assume that the basic mesh is cubic so that the Laplace algorithm (2.4) becomes simply

$$V(x,y,z) = \tfrac{1}{6}\big(V(x+\Delta x,y,z) + V(x-\Delta x,y,z) + V(x,y+\Delta y,z)$$
$$+V(x,y-\Delta y,z) + V(x,y,z+\Delta z) + V(x,y,z-\Delta z)\big) \quad (6.15)$$

or written for the array $V(I,J,K)$

$$V(I,J,K) = \tfrac{1}{6}\big(V(I+1,J,K) + V(I-1,J,K) + V(I,J+1,K)$$
$$+V(I,J-1,K) + V(I,J,K+1) + V(I,J,K-1)\big) . \quad (6.16)$$

In the yz-plane where $I = 0$ one uses the fact that $V(-x,y,z) \equiv V(x,y,z)$ and the algorithm

$$V(0,J,K) = \tfrac{1}{6}\big(2V(1,J,K) + V(0,J+1,K) + V(0,J-1,K)$$
$$+V(0,J,K+1) + V(0,J,K-1)\big) \quad (6.17)$$

is obtained. The extensions to the xz-plane and the optic axis are obvious.

Unlike the case of cylindrical deflectors, we now want to define lenses with arbitrary shape; it is convenient to introduce an array separate from the potential array for this purpose. The non-zero elements of this new array $Vl(I,J,K)$ are either a very small potential of say 10^{-8} V or the unit potential of 1 V (Sect. 2.1), while $Vl(I,J,K) \equiv 0$ for all points in space not covered by metal electrodes. A simple reading code can then be used to make plots of the lens elements in arbitrary cross-sections. Such plots are very useful, almost essential indeed, to make sure that one has indeed defined the lenses exactly as they were meant to be. The Laplace algorithm is then designed to jump to the next I,J,K unit when one hits a lens element, i.e. when $Vl(I,J,K) \neq 0$. The potential at that particular position $V(I,J,K)$ is set equal to $Vl(I,J,K)$. The outer boundary of the array must be excluded from the algorithm. There the potential $V(*)$ is set equal to $Vl(*)$. The Laplace equation must be solved as many times as there are independently variable potentials on lens elements; the unit potential of 1 V is applied to each element in turn, while the other elements of the lens are held at the negligible small potential of 10^{-8} V. Since the lens elements occupy a

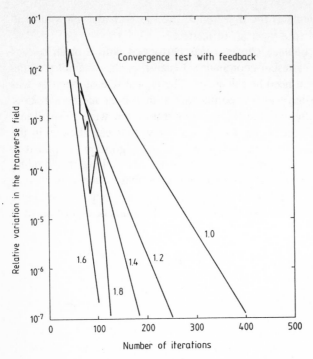

Fig. 6.1. Relative variation in the transverse electric field vs the number of iterations. The data refer to a cathode emission system with oval lenses. The parameter is the feedback-factor as defined in (2.4). For a feedback-factor of 1 (i.e. no feedback) the convergence is rather slow. For larger feedback-factors, the convergence becomes more and more rapid until the results start to oscillate. The optimum feedback-factor is between 1.6 and 1.8 for the lens system here

much larger volume than the space in which the potential is actually needed to run the trajectories, only the centre part of the converged potential need to be stored. It is advisable also to store all key parameters such as array dimensions, the key dimensions of lens elements, and the cathode and target positions and let the trajectory program later read and adjust itself to these parameters. If a calculation of the space charge potential is attempted, one must also store the array $Vl(*)$ defining the lenses in order to let the algorithm solving the Poisson equation take the metallic electrodes into account properly.

The convergence of the Laplace algorithm is speeded up substantially by positive feedback (2.5). The optimum choice for the feedback parameter was found experimentally by testing the convergence in the field as a function of the feedback parameter. In Fig. 6.1 the relative variation of the transverse electric field at a particular point on the optic axis is plotted as a function of the number of times the Laplace algorithm has run over the entire array. One sees that after a starting-up period, the field converges exponentially with the number of cycles, yet with a rather slow decay rate for a feedback-factor of one [which means no positive feedback (2.5)]. For higher feedback factors the convergence becomes faster and faster until the field begins to oscillate with the number of iterations.

92

Eventually the oscillations become too large and the procedure is unstable. For the lens elements in the emission system, a feedback of 1.6 was used, which made the convergence rather rapid. By exploiting the fact that the deviation in the field starts to oscillate when the feedback becomes too high, it is straightforward to let the feedback automatically adjust to an optimum value. The iterations were stopped when the relative variation in the *field* after each iteration was below 10^{-6}. For the emission system with five independently variable potentials, an entire lens system could be calculated in about 1 h on the 320 series Hewlett-Packard with an Infotek Basic 5.1 compiler.

We now describe a few key features of the trajectory program where this differs from the one used for cylindrical deflectors. Here, electron trajectories need to be calculated only near the optic axis. This can be exploited to speed up the calculation of the electric field at each instantaneous electron position by performing a major fraction of this calculation before the integration of the trajectories. For example, one may determine the transverse electric field for all values of the index K counting along the optic axis, which is referred to as the z-axis in this and the next chapter, dealing with lens systems. We take the transverse x-component of the field as an example and expand the potential around the optic axis as in (2.10)

$$V(I, J, K) = a_0(J, K) + a_1(J, K)(\Delta x)^2 I^2 + a_2(J, K)(\Delta x)^4 I^4 \dots . \qquad (6.18)$$

Because of the mirror symmetry, with the yz-plane as a mirror plane, only the even terms in the expansion survive. After defining a matrix $\tilde{a}$ with elements $\tilde{a}_i$

$$\tilde{a}_i(J, K) = a_i(J, K)(\Delta x)^{2I} \qquad (6.19)$$

one obtains $\tilde{a}$ by the operation

$$\tilde{a}(J, K) = M^{-1} \cdot V(J, K) \qquad (6.20)$$

for each pair (J, K), provided M has the elements

$$M_{ij} = (i)^{2j} . \qquad (6.21)$$

The quantity 0^0 is set equal to unity. In some languages, the operation 0^0 results in an error message, which must then be taken into account when the matrix M_{ij} is defined. The x-component of the field is then obtained as

$$\mathcal{E}_x(J, K) = 2a_1(J, K)x + 4a_2(J, K)x^3 \dots . \qquad (6.22)$$

The matrix $a(J, K)$ and a similarly obtained matrix $b(I, K)$ representing the y-component of the field depend on the applied potentials, not on the instantaneous position of the electron. The matrices can therefore be calculated once the applied potentials have been defined. The only equation that needs to be solved for the each position of the electron is (6.22). For positions of the electron between the grid points (J, K); $(J + 1, K)$; $(J, K + 1)$; $(J + 1, K + 1)$ one may use a linear interpolation. For the lens system we have studied it was sufficient to take (6.22) up to the third order. A fifth order expansion was actually used in the program to be on the safe side. We note that with Eq. (6.22) one is no longer confined to the positive quadrant. The equation applies also to the negative values of x, despite the fact that the potential array is only defined in the first positive quadrant. Thus the program automatically integrates trajectories in the entire lens with no further precautions.

For the z-component of the electric field we used the second-order expansion

$$V(I, J, z) = c_0(I, J) + c_1(I, J)(z - z_K) + c_2(I, J)(z - z_K)^2 , \qquad (6.23)$$

where z_K is the instantaneous z-position of the electron at the last integer point K defined by

$$K = \text{INT}(z/\Delta z) , \quad z_K = \Delta z K . \qquad (6.24)$$

The field is then calculated again by the matrix inversion method described above, for each integration step in the trajectory calculation. The z-component of the electric field varies in second order with the deviation from the optic axis. For positions x, y off the optic axis, one may interpolate between the grid points I, J using a proper expansion of the coefficients $c_i(I, J)$ as even powers of I, J. For the rays near the axis as used here, we found however the results of such a more elaborate integration procedure to be practically identical with the results obtained with $V(I, J, z) \equiv V(0, 0, z)$ in (6.23). The z-component of the electric field can therefore safely be calculated from the potential on the optic axis. This is not surprising. We recall that in the gaussian optics of paraxial rays for cylindrical lenses all the imaging properties are determined by the potential on the axis [6.5, 6].

The trajectories were calculated using the scheme

$$\dot{x}(t + \Delta t) = \dot{x}(t) + \mathcal{E}_x(x, y, z)\Delta t ,$$
$$x(t + \Delta t) = x(t) + \dot{x}(t)\Delta t + \tfrac{1}{2}\mathcal{E}_x(x, y, z)\Delta t^2 , \qquad (6.25)$$

as described earlier with the corresponding equations for the y- and z-components. Changing to a more elaborate integration scheme such as the Runge-Kutta procedure [6.7] did not reduce the computing time, presumably because in our case the field between the grid points is not known exactly and is calculated by interpolation.

The convergence of the integration was tested as a function of the time steps Δt. As a measure of Δt we use the dimensionless quantity $\Delta z/v_{\text{slit}}\Delta t$, where v_{slit} is the velocity at the target. In Fig. 6.2 the angular aberration for a particular emission lens system and the error in the kinetic energy of the electrons at the target are plotted as a function of $\Delta z/v_{\text{slit}}\Delta t$. The angular aberration is here defined as the difference between the foci for electrons emerging from the cathode tip with angles of 5° and 40°, respectively. This quantity converges to a finite value. One sees that the results are sufficiently converged for values of $\Delta z/v_{\text{slit}}\Delta t$ of about 20. The remaining error in the focal length corresponds to a very small deviation in the lateral x position of about 0.01 mm because of

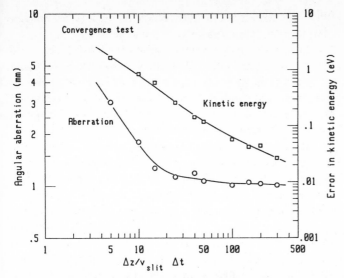

Fig. 6.2. Angular aberration and error in the kinetic energy at the target as a function of the inverse time step in the trajectory integration. The converged result for the geometrical parameters of the trajectories is reached when $\Delta z/V_{\text{slit}}\Delta t$ is above ~ 20 with a potential of $V_{\text{slit}} = 10\,\text{eV}$ at the target. This corresponds to about 150 integration steps for each trajectory. The error in the kinetic energy of the beam at the target falls rather slowly with the number of integration steps. The kinetic energy can however be calculated directly from the initial energy and energy conservation

the small angular divergence of the beam at the target. Similarly all other beam parameters related to the geometry of the trajectories converge rapidly. The error in the kinetic energy at the target disappears only as Δt. When the kinetic energy is of interest, one should make use of energy conservation directly rather than calculating the energy from the integration.

The integration of one trajectory with an arbitrary emission angle at the cathode takes about 1 s. In the determination of the foci in the two symmetry planes of the lenses one may use simplified trajectory subroutines and consider the x, z, or y, z components of the electric field only, which leads to a further gain in speed. Once a set of potentials has been determined leading to an approximate focus near the target a search code is useful, which keeps changing one key potential in a systematic manner, until the focus coincides with the target with a predetermined precision. In order to determine the essential beam parameters such as the input current into the monochromator and the brightness at the input slit, the Maxwellian energy distribution of the electrons emitted from the cathode has to be taken into account. A Maxwellian distribution $E\exp(-E/k_{\text{B}}T)$ of the energies E is generated by first selecting a random value of E_0 between zero and a maximum value of about ten times the particular energy where $E\exp(-E/k_{\text{B}}T)$ has its maximum. One then generates a second random number between zero and the maximum value of the function $E\exp(-E/k_{\text{B}}T)$. If this number is larger than $E_0\exp(-E_0/k_{\text{B}}T)$ a next trial value of E_0 is selected, otherwise E_0 is used to start a trajectory. While this code is not the most efficient one, it is fast enough not to be a time-limiting factor in the integration of a bundle of trajectories.

The space charge and the space charge potential were calculated as described earlier in Sect. 5.1. As we shall see later with specific examples, a sufficient brightness at the entrance slit of the monochromator and a sufficient total current are attained with cathode emission currents corresponding to comparatively low space charge densities. This allows us to investigate the effect of space charge on the trajectories and the relevant beam parameters in a first order approximation, in which the space charge is obtained from a bundle of trajectories traversing the system without the presence of space charge. In other words, one is still far from the conditions in which a virtual cathode is formed near the cathode tip and the total current becomes space charge saturated. That this is indeed so for a particular emission system may be tested in two ways. Firstly, one may rise the space charge potential calculated to first order by multiplication with a higher current until trajectories emitted from the cathode are repelled from the space charge barrier formed near the cathode. This current should be much larger than the current for which the first order space charge potential is used. Secondly, one may iterate the space charge calculation for a particular current of interest by calculating the space charge and space charge potential with a bundle of trajectories in the presence of the first-order space charge potential, and so forth. Just as for the cylindrical deflector, we found that the first-order space charge potential was sufficiently well converged for the emission currents of interest here. It should be however mentioned that while the calculated first-order properties of emission systems – the angular apertures and the achievable brightness – agreed with experimental measurements, the energy distribution of the electrons entering the monochromator was narrower than calculated. This may be related to the fact that the shape of the cathode tip cannot be modelled satisfactorily with the relatively coarse mesh.

6.3 Three Different Emission Systems

In electron energy loss spectroscopy, repeller-type cathode emission systems have enjoyed rather widespread use, while on the other hand guns with Wehnelt cylinders are typically employed in applications where a beam of small angular divergence and diameter at high electron energy is needed. So far as we know the use of repeller cathodes was introduced by *Ehrhardt* and collaborators [6.8] in the 1960s. It was not stated whether the choice was made empirically or on theoretical grounds. We have tested two different small Wehnelt-type cathodes and found their properties to be far inferior to repeller cathodes for the application here. We assume that this result comes about because of the negative potential applied to the Wehnelt cylinder. This negative potential is necessary in order to obtain a focus. As a consequence, electrons near the cathode travel in a region of smaller potential. This is clearly visible in a plot of the space charge density vs

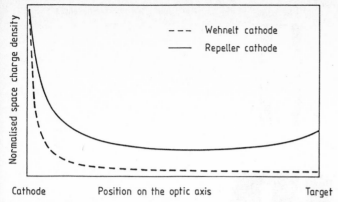

Fig. 6.3. Distribution of the space charge density along the optic axis for a Wehnelt emission system and a repeller emission system plotted as dashed and full lines, respectively. The data refer to the same target potential but are normalised to the maximum space charge density at the cathode for each system. The space charge density of the Wehnelt emission system is more localised in the vicinity of the cathode and is actually also much higher there for comparable dimensions. Consequently the Wehnelt cathode saturates at lower emission currents

the distance from the cathode. For the Wehnelt-type emission system the space charge density (dashed line in Fig. 6.3) is more concentrated in the region near the cathode while, for the circular repeller system to be discussed later in more detail, the space charge is more evenly distributed, and consequently the space charge density is lower on the whole for the repeller emission system. Further investigation therefore focuses on the optimisation of repeller cathodes. Guided by the slit geometry of cylindrical deflectors used in this laboratory, a repeller-type emission system with lens elements having slit-shaped cross sections is investigated. Obviously such an emission system can be constructed so that one forms a horizontal image at the entrance slit, while one may have a nearly parallel beam in the vertical plane. A slit of narrow width and large height can hence be more or less illuminated. It was widely believed for years that filling the entrance slit in this manner is the only effective way of operating a monochromator. We have already seen, in the investigation of space charge effects in the cylindrical deflector, that the height up to which a slit of 0.15–0.3 mm width should be filled should be limited to about 1–2 mm. It was one of our major surprises in this study that, at least for an ideal point source cathode, an emission system that has circular symmetry around the optic axis provides rather good results as well.

We have studied the optical properties of three classes of repeller emission systems in greater detail, one with slot lenses, one with oval lenses, and one with circular lenses. In each of these classes we have varied many of the parameters characterising the dimensions of the lens elements and the distances. The specific system discussed below represent a relative optimum with regard to the beam parameters demanded by a pre-monochromator operating with pass energies larger than ~ 1.5 eV. For very low pass energies the detailed investigation of experimentally realized spectrometers suggested that emission systems providing

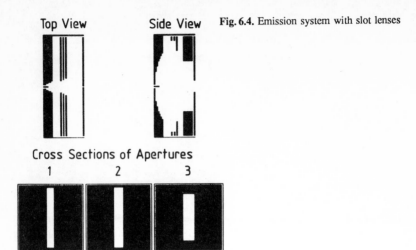

Top View   Side View   **Fig. 6.4.** Emission system with slot lenses

Cross Sections of Apertures
1   2   3

a beam of a smaller angular aperture at the entrance slit of the monochromator would be advantageous when resolutions below 2 meV are attempted (Chap. 8).

We now describe the geometry of different lenses in detail with the help of computer-generated drawings of the cross-sections. The cross-sectional views are generated by drawing a filled square around each point were a nonzero value of the lens array $Vl(*)$ has been declared. Thus the drawings are accurate to 0.25 mm. As we have mentioned above, differences in dimensions of this order are not important anyway. We first comment on the slot lens system depicted in Fig. 6.4. The system consists of a repeller, three slot apertures plus a horizontal bar (see side view) and the entrance slit of the monochromator, which is the target for the beam emerging from the cathode. The cathode is drawn as a little tip in the left center of the top view and the side view. The repeller potential also surrounds the entire system. Because of the negative potential applied to the repeller, the electrons are kept within the emission system and no stray electrons can reach the electron detector at the end of the spectrometer. This is of some importance since one produces about 10^{13} electrons/s, whereas the dark count rate of the detector is less than 1 Hz. The repeller has a different curvature in the two cross sections. These curved shapes are obtained by using a milling wheel of 50 mm diameter. Five independently variable potentials can be applied to the lens system, one to the repeller, one to each aperture, and one to the target aperture. The latter determines the input energy for the monochromator. In Sect. 5.2 we have calculated the example of a pre-monochromator for a kinetic energy at the entrance aperture of 1.5 eV. In keeping with this example we calculate the cathode systems with 1.5 V again applied to the entrance slit of the monochromator. The extra bar shown in the side view of Fig. 6.4 is electrically connected to the last lens aperture. Held at a negative potential, the bar provides for an independently adjustable vertical focus, as we shall see.

Cross-sections of the second class of cathode systems are shown in Fig. 6.5. The slot apertures are replaced by apertures with oval openings having a height-to-width ratio of 2. The extra correcting bar of the first lens is removed because it was assumed that sufficient focusing in the vertical plane (plane of the side view) would be obtained by closing the slots to ovals. The repeller remained the same as for the slot aperture system.

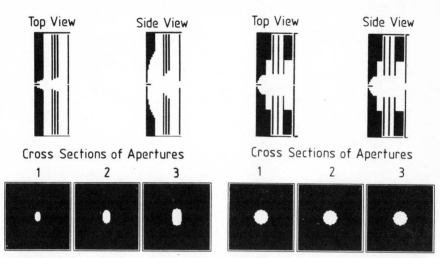

Fig. 6.5. Emission system with oval lenses   **Fig. 6.6.** Emission system with circular lenses

The third system (Fig. 6.6) is the simplest of all, containing only rotationally symmetric elements, apart for the entrance slit, which could of course be replaced also by a round aperture. We recall however that the slit width determines the resolution of the monochromator and must be matched to the design of the monochromators and analysers, while the height of the slit can be varied over a large range without affecting the resolution. When the disk of confusion formed by the trajectories of electrons having the monochromator pass energy is larger than the slit width, then a slit height larger than the width provides more useful feed current into the monochromator. Interestingly, the slit form of the target aperture, the only non-circular aperture in the emission system, has a small but noticeable asymmetry effect on the trajectories. Like the apertures, the repeller is now also circularly symmetric. The shape of the repeller is obtained by grinding a hemisphere of radius 4 mm into the repeller plate. (The shaping of the elements in the computer code actually mimics the action of a milling cutter.) The centre of the sphere is situated at a point 1 mm to the right of the inner plane of the repeller (Fig. 6.6). The cathode tip is placed 2.5 mm to the left of the centre of the sphere. The emission system with circular lenses also includes two screening electrodes, connected to the repeller and the third aperture. The purpose of these electrodes is to hide the ceramic insulators needed to stack the lens elements and fix their position. Electrons on the trajectories "see" the creamic beads though a

Table 6.2. Essential design parameters of three emission systems. The symbols z_i, h_i and w_i denote the positions on the optic axis, the height, and the width of the ith aperture, respectively, in mm. Each system was actually tested in a spectrometer

Type of emission system	Slot	Oval	Circular
Cathode z-position	1.5	1.5	1.5
Cathode surface area	0.01×0.01	0.01×0.01	0.01×0.01
z-position of inner boundary of repeller	3.0	3.0	3.0
z_1	6.5	6.5	6.0
h_1	36.0	6.0	8.0
w_1	5.0	3.0	8.0
z_2	7.5	7.5	8.0
h_2	36.0	8.0	8.0
w_2	6.0	4.0	8.0
z_3	8.5	8.5	10.0
h_3	28.0	10.0	8.0
w_3	7.0	5.0	8.0
z_{slit}	15.0	12.5	14.0
Height of correction bar	20.0	–	–
Diameter of screening electrode	–	–	18.0

narrow gap between metallic electrodes, which screens the possibly high potential of charged insulating surfaces. The key parameters of all three emission systems are summarised in Table 6.2.

Once the emission systems have been defined and the Laplace equation has been solved, one may begin to explore feasable conditions for obtaining a horizontal focus at the entrance slit. We also remember that a small angular divergence of the beam with respect to both the horizontal plane and the vertical plane is desirable. In keeping with the definitions in the cylindrical deflector, we use α for the angle with the optic axis in the horizontal plane and β in the vertical plane. Once the target potential has been fixed, one still has four independently variable potentials. Some guidance to suitable subspaces of the four-dimensional parameter space is provided by the following description of the main effects of the individual potentials. We take the slot lens as an example, since there the elements have the most clearly distinguishable function.

The lens operates as an accelerating-retarding lens. A comparatively high potential is applied to the centre aperture to accelerate the electrons rapidly as they leave the cathode and to remove them from the critical region where a space charge cloud and a virtual cathode may be formed. The magnitude of this potential is rather arbitrary, the upper limit basically being a matter of convenience. The gain in performance becomes rather sluggish at high voltages. When the potential of the centre aperture is varied, one needs to scale the other potentials too, of course. The repeller is always kept at a negative bias with respect to the cathode in order to "repel" the electrons. The bias determines rather critically the horizontal focus of the beam. Some focusing in the vertical plane is also achieved by the repeller but much less than in the horizontal plane, because of the smaller curvature of the repeller in the vertical plane. The first aperture has a potential typically between those of the repeller and the second aperture, providing some Wehnelt-type focusing action. The third aperture is at a negative potential. The main effect of this potential is to shape the beam in the vertical plane into a converging beam and to provide a vertical focus between about three times the slit distance and infinity. Needles to say, all potentials are interdependent. If one changes the potential on aperture 3, say, not only does the vertical focus change but also the horizontal. The last adjustment is always performed with the repeller, which critically determines the horizontal focus, while the remaining parameters of the beam remain unchanged, to first order. For the repeller potential, one can therefore easily write a subroutine for an automatic search for the focus.

6.4 Electron Optical Properties
of the Three Different Emission Systems

We begin the discussion of the electron optical properties with the presentation of a set of trajectories in the vertical and horizontal planes, the potential distribution along the optic axis, and the image of the cathode at the target. We show these results for zero emission current and for an emission current of $1\,\mu A$ (Figs. 6.7–13) in order to elucidate the effect of space charge on the trajectories and the potential. The cathode is assumed to be an ideal point source for the moment. The potentials on the lens elements are adjusted to have a focus at the target in the horizontal plane for a pair of beams leaving the cathode at an emission angle $\alpha = \pm 20°$, whereas for the vertical plane we attempted to have a nearly parallel beam in the emission systems with slot and oval lenses. There is obviously some arbitrariness involved here, and it is useful to study the emission systems with different focusing in order to build up experience about their performance under different conditions. The final choice of the potentials was made in keeping with the set of optimised potentials obtained in experimental tests of the cathode systems. In the calculation we kept the potentials at the same value for high and low emission currents, although readjustment of the potentials slightly improves the currents and brightness achievable. We have not readjusted the potentials since this would have been rather difficult to perform in an unambiguous manner.

The results for the slot lens emission systems are shown in Fig. 6.7. Relatively large angular aberrations are seen in the horizontal plane and also in the vertical plane. In the vertical plane the absence of angular aberrations would result in a parallel beam for all emission angles β since the rays are emitted from an ideal point source. The potential along the axis (Fig. 6.7c) has a minimum near the target because of the negative potential $V_3 = -13\,V$ applied to the third aperture and the bar, which is needed to shape the beam in the vertical plane. It is obvious that this minimum can have unpleasant effects on the beam parameters when it

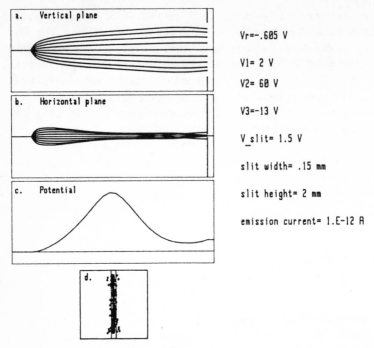

Fig. 6.7. Trajectories of electrons (**a**) in the vertical plane and (**b**) in the horizontal plane for the emission system with slot lenses. (**c**) The potential along the axis. (**d**) The positions of the electrons at the target slit, the dimensions of which are 0.15 mm × 2 mm. The trajectories refer to electrons emitted from a point cathode at an energy equivalent to the maximum of a Maxwellian flux distribution with a temperature $T = 2000$ K

is further deepened by the space charge. The arrival points of 200 electrons at the target with initial emission angles randomly distributed between $\alpha = \pm 40°$, $\beta = \pm 40°$ are shown in Fig. 6.7d. The initial energy of the electrons was held at the maximum of the Maxwellian flux distribution. One sees that the slit is nicely filled in both width and height. Thus a transmission of nearly 100% is achieved for electrons having this energy when emerging from a point source. The latter caveat is essential, because the three lens systems have a relatively large magnification (4.5 and 7.5, respectively). Figure 6.8 displays the same results, but now for an emission current of $1\,\mu$A. The effect on the trajectories, in particular near the cathode, is clearly visible. Raising the current further soon results in the build up of a virtual cathode near the emission tip, a regime of emission currents where our approch to the space charge problem no longer applies. The spreading of the beam at the target due to the space charge is obvious when one compares Fig. 6.7d and 6.8d. We recall that this spreading is largely due to increased angular aberrations and can therefore not be reduced substantially by applying slightly modified potentials.

The results for the oval lenses are depicted in Figs. 6.9 and 6.10. Again the slit is nicely filled for electrons with an emission energy at the maximum of the

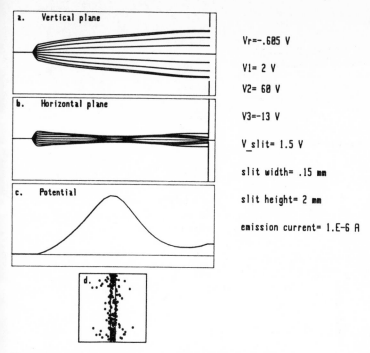

a. Vertical plane

b. Horizontal plane

c. Potential

d.

Vr=-.685 V

V1= 2 V

V2= 68 V

V3=-13 V

V_slit= 1.5 V

slit width= .15 mm

slit height= 2 mm

emission current= 1.E-6 A

Fig. 6.8a–d. The same as Fig. 6.7 for an emission current of 1 μA

Maxwellian. In contrast to the slot system, electrons diverge in the vertical plane (Fig. 6.9a) even in the limit of zero emission current and diverge even more for larger currents (Fig. 6.10a). This is due to the absence of the extra bar in the emission system, which was used to shape the beam in the slot lens system (compare Figs. 6.4 and 6.5). Thus with the oval lens system studied here, the beam shaping in the vertical direction remains unsatisfactory. We assume that by modifying the dimensions of the lens elements one could eventually obtain an emission system employing ovally shaped lenses only and still achieve the better beam shaping of the slot lens system equipped with the horizontal electrode bar. We did not explore this possibility in greater detail since we expected to arrive eventually at a system which, though easier to manufacture, would provide electron optical properties rather similar to those of the slot system.

The final system considered here is the emission system with circular lenses (Figs. 6.11 and 6.12). In contrast to the two former ones, this system can be operated with a rather large field near the cathode, which is apparent from the steep increase of the potential near the cathode. This is a natural feature of the circular system, where no attempt is made to focus the beam in one plane and spread it in the other in order to fill a slit. In fact, a proper focus could be obtained with even higher potentials on the repeller and first lens elements. Generally speaking, the system is rather good natured because it tolerates quite different potentials and still provides for a good focus at the target. This is a

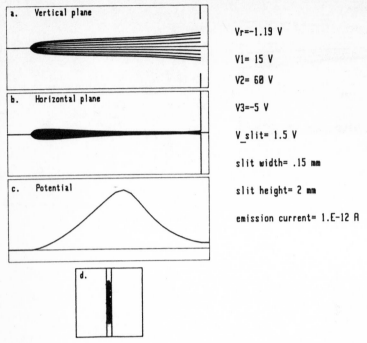

Fig. 6.9. Trajectories of electrons (**a**) in the vertical plane and (**b**) in the horizontal plane for the emission system with oval lenses. (**c**) The potential along the axis. (**d**) The positions of the electrons at the target slit, the dimensions of which are 0.15 mm × 2 mm. The trajectories refer to electrons emitted from a point cathode at an energy equivalent to the maximum of a Maxwellian flux distribution with a temperature $T = 2000\,\mathrm{K}$

consequence of the fact that the system is now redundant in the number of apertures involved. We have retained this redundance since two of the apertures are split along the vertical plane and one along the horizontal plane. This allows the beam to be directed into the entrance slit even in the presence of asymmetries in the field, which may arise from inhomogeneous charging or residual magnetic fields due to the current-bearing leads to the cathode. As a consequence of the higher electric fields near the cathode tip, the effect of increasing the current has much less effect on the shape of the beam. The extra broadening seen in Fig. 6.12d compared to 6.11d can be reduced further by applying a somewhat different potential as the disk of least confusion does not fall at the target for $1\,\mu\mathrm{A}$ emission current (Fig. 6.12a, b). With the set of potentials used in Figs. 6.11 and 6.12, the system performs best when the cathode is nearly a point source.

Fig. 6.11. Trajectories of electrons (**a**) in the vertical plane and (**b**) in the horizontal plane for the emission system with circular lenses. (**c**) The potential along the axis. (**d**) The positions of the electrons at the target slit, the dimensions of which are 0.15 mm × 2 mm. The trajectories refer to electrons emitted from a point cathode equivalent to the maximum of a Maxwellian flux distribution with a temperature of $T = 2000\,\mathrm{K}$

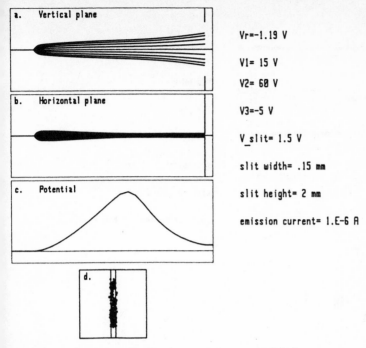

Vr=-1.19 V

Vl= 15 V

V2= 60 V

V3=-5 V

V_slit= 1.5 V

slit width= .15 mm

slit height= 2 mm

emission current= 1.E-6 A

Fig. 6.10a–d. The same as Fig. 6.9 for an emission current of 1 μA

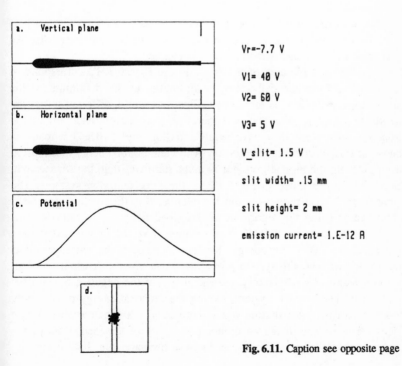

Vr=-7.7 V

Vl= 40 V

V2= 60 V

V3= 5 V

V_slit= 1.5 V

slit width= .15 mm

slit height= 2 mm

emission current= 1.E-12 A

Fig. 6.11. Caption see opposite page

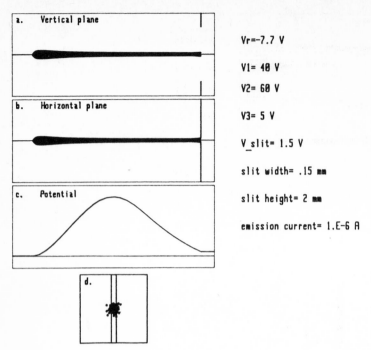

a. Vertical plane

Vr=-7.7 V

Vl= 48 V

V2= 68 V

b. Horizontal plane

V3= 5 V

V_slit= 1.5 V

slit width= .15 mm

c. Potential

slit height= 2 mm

emission current= 1.E-6 A

d.

Fig. 6.12a–d. The same as Fig. 6.11 for an emission current of 1 μA

For larger cathode areas, a set of potentials with a negative bias on the last lens aperture as used for the slot lens and the oval lens provides better results. We shall come back to this issue at the end of this chapter.

The trajectories and pictures of the focus in the Figs. 6.7–12 correspond to electrons emerging from the cathode with an energy at the maximum of the Maxwellian flux distribution, as mentioned. Electrons with a different initial energy have different trajectories and will therefore pass the entrance slit of the monochromator only when they have small emission angles. The net result is that, because of this "chromatic" aberration of the lenses, the energy distribution of the beam entering the monochromator is much narrower than the Maxwellian distribution. This is illustrated in Fig. 6.13 for the three emission systems. The broad curve in each case represents the Maxwellian flux distribution generated with 20 000 beams using the algorithm as described before. The cathode temperature was assumed to be 2000 K. The resulting FWHM of the distribution is in keeping with (6.10). The energy distribution passing the entrance slit is indeed narrower, with the slot system providing the narrowest distribution of all the emission systems. This is largely a consequence of the negative bias on the last lens element. The circular system, having the broadest energy distribution, also provides a narrower distribution when operated with a negative potential on the last lens element. The increased monochromaticity of the feed beam of the monochromator was already taken into account in the parameter listing specify-

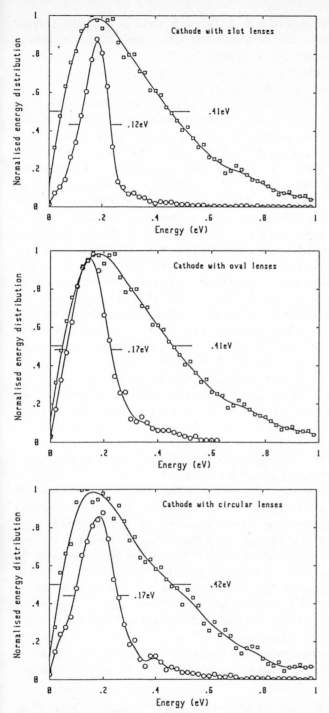

Fig. 6.13. The Maxwellian flux distribution as a function of energy (□) is compared to the energy distribution of electrons actually passing the slit for the three different emission systems

ing the pre-monochromator (Table 5.1). This may serve as one more example that, in the overall optimisation of spectrometers, one needs a priori knowledge of the basic properties of each component. Needless to say, the width of the transmitted energy distribution depends on the dimension of the entrance slit of the monochromator and the potential on the slit.

The differences between the three lenses are much larger for the other beam parameters of interest. One of these beam parameters is clearly the input current, that is, the fraction of the emission current that can pass through the slit. The investigation of a pre-monochromator in 5.2 told us that the entrance current should be 50 nA for 1.5 eV pass energy. This value, which refers to a beam having no divergence with respect to the angle β and a divergence in α by $\pm 1.5°$, may be considered as a lower bound. A considerable fraction of the beam actually produced by the cathode emission system has larger angles α and also the divergence in β can be as high as 5° for the oval lenses. We can thus assume that, with a feed beam of this type, the monochromator will tolerate higher input currents of several hundred nA at 1.5 eV. We have also learned however that the monochromator will not transmit electrons unless their aperture angles are small enough, typically $|\alpha_{1m}| < 1.5°$ and $|\beta_{1m}|$ less than about 1°. Furthermore, the energy of the electrons should fall within the interval of transmitted electrons. A useful quantity is therefore the current per energy interval (taken at the maximum brightness of the cathode) that falls within a small range of entrance angles, which we take to be $|\alpha_{1m}| < 1.5°$, $|\beta_{1m}| < 1°$. We refer to this current as the "radiance". The radiance is not a universibly definable quantity as the brightness is, but serves a useful purpose here in the comparison of different emission systems. In Fig. 6.14 the input current and the radiance as defined above are plotted as a function of the cathode emission current. Each point is calculated with 1000 trajectories. For the slot lens, the input current levels off and the radiance passes through a maximum. This is a consequence of the fact that the cathode operates close to the point where a virtual cathode is formed.

The emission system with oval lenses yields a rather poor radiance. In fact the radiance is not high enough to feed the system of pre-monochromator and monochromator discussed in Sects. 5.1 and 5.2. We can see this by considering the following example. Let us suppose that we can feed the pre-monochromator with a maximum current of 100 nA which is already twice the value suggested in Fig. 5.13. Then the radiance is ~ 0.06 nA/meV. According to Fig. 5.13 the resolution of the pre-monochromator is about 9 meV. With a radiance of ~ 0.06 nA/meV, we calculate a maximum output current of the pre-monochromator of ~ 0.5 nA, a result which assumes that there are no transmission losses. The actual output current that the pre-monochromator can supply is 2 nA (Fig. 5.13), provided the monochromator is fed with a beam of sufficient radiance. This example demonstrates the usefulness of the quantity radiance. For

Fig. 6.14. Input current per unit energy in a small angle $|\alpha_{1m}| < 1.5°$, $|\beta_{1m}| < 1°$ ("radiance") and the total input current vs the emission current for the three emission systems when the latter are equipped with a point cathode. The data points are calculated with 1000 trajectories each

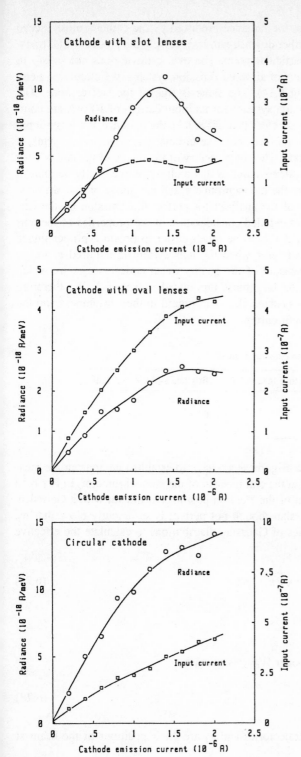

Fig. 6.14. Caption see opposite page

a given spectrometer design, the radiance produced by the cathode must exceed the radiance at any point further downstream by precisely the product of all transmission losses up to that particular point. The oval cathode does not satisfy to this requirement. The simplest of all three emission systems, the circular system, however does satisfy it (Fig. 6.14). The same is true of the slot cathode. The radiances for the three emission systems for an input current of 100 nA are compared in Table 6.3. It is already clear that, if we take the radiance at a given input current as a measure of the performance, the circular emission system wins, if one has a point source cathode. The slit geometry of the image of the cathode for both the slot and the oval systems requires that a slit must eventually be imaged by the lens system following the monochromator and the image of that slit must fall within the entrance slit of the analyser for perfect transmission by the lens system. Even without further detailed consideration of the properties of the lens systems, we appreciate that it should be much easier to obtain a proper image at the entrance slit of the analyser when the lens system is required to image a small square object. A measure of the quality of an emission system, which emphasizes this aspect, is the brightness introduced in Sect. 6.1. We therefore compare the three emission systems also with regard to their brightness and the ratio of the brightness to input current.

Table 6.3. Performance of the three emission system

Lens type	I_{in} [nA]	Radiance [nA/meV]	Brightness [A/eV cm² rad]
Slot	100	0.28	0.3
Oval	100	0.06	0.09
Circular	100	0.38	3

In order to calculate the brightness we have to establish an appropriate measure of the size of the beam at the entrance slit of the monochromator. In Fig. 6.15 we compare the distribution of the electron positions at the slit with a Gaussian. The resemblance to a Gaussian though not perfect is sufficiently close that we can use the known properties of Gaussian distributions to calculate the effective width of the beam. With

$$\int_{-\infty}^{\infty} \exp(-x^2/2\sigma^2) = \sigma\sqrt{2\pi} \quad \text{and} \tag{6.26}$$

$$\sigma^2 = \frac{1}{n-1} \sum_{i=1}^{n} x_i^2 \tag{6.27}$$

the effective width of the beam $W_{\text{eff } x}$ is given by

$$W_{\text{eff } x}^2 = \frac{2\pi}{n-1} \sum_{i=1}^{n} x_i^2, \tag{6.28}$$

where n is the number of trajectories and x_i are the x positions of the beam at

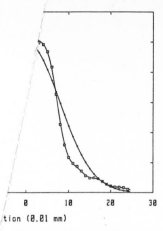

tion (0.01 mm)

as a function of the x-position at the entrance aperture compared with
a gaussian ... 26–28)

the target measured from the optic axis. The same equation is used for the width
in the y-direction and the width of the angular distributions in α and β.

By using $W_{\mathrm{eff}\,x}$ and the corresponding quantities $W_{\mathrm{eff}\,y}$, $W_{\mathrm{eff}\,\alpha}$, $W_{\mathrm{eff}\,\beta}$ for y,
α, and β respectively we calculated the brightness with the help of (6.9):

$$B = \frac{I_{\mathrm{emission}}T(E_{\mathrm{m}})}{ek_{\mathrm{B}}TW_{\mathrm{eff}\,x}W_{\mathrm{eff}\,y}W_{\mathrm{eff}\,\alpha}W_{\mathrm{eff}\,\beta}}\,. \tag{6.29}$$

Here I_{emission} is the total emission current and $T(E_{\mathrm{m}})$ is the fraction of the electrons with an energy at the maximum of the Maxwellian distribution that can
pass through the entrance slit of the monochromator.

The resulting values for the brightness as a function of the emission current
are shown in Fig. 6.16. The best emission system is obviously the one which provides the highest brightness for the lowest input current. The ratio of brightness
to input current is therefore also plotted in Fig. 6.16. The highest value of this
ratio is obtained with the circular emission system. The circular emission system
is also the only one for which the brightness comes near to the maximum brightness of a tungsten emitter (Table 6.1). We note further that on general grounds
the maximum brightness/input current is expected for small emission currents.
The presence of the maximum for the slot lens near $1\ \mu$A emision current indicates that we do not have the optimum focusing conditions for small emission
currents. This is in fact visible in Figs. 6.7 and 6.8. For zero emission current we
have a slightly converging beam in the vertical plane (Fig. 6.7a) while the beam
becomes more parallel with higher currents (Fig. 6.8a), which results in a higher
brightness/input current for the higher emission current. By reducing the negative bias on the bar electrode one may shift the maximum in the brightness/input
current in Fig. 6.16 (top) to smaller emission currents.

In conclusion, it appears that the circular emission system performs best
and is also easiest to construct and to operate. Alas, this is true only when the

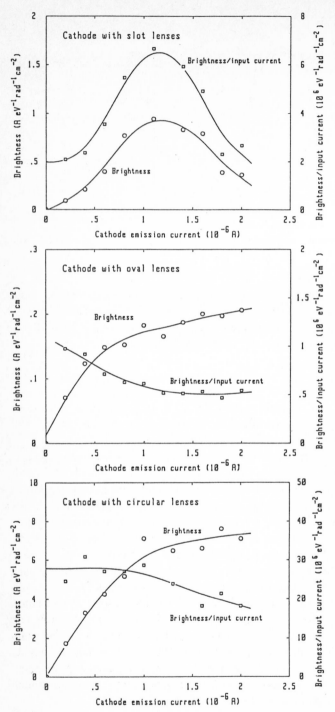

Fig. 6.16. Brightness and brightness/input current vs the emission current for the three different emission systems when equipped with a point cathode

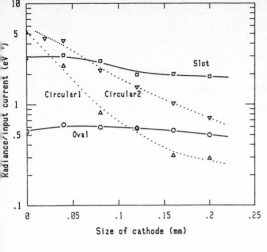

Fig. 6.17. Brightness/input current for the three emission systems as a function of the length of the squared area of the cathode. The second data set for the circular cathode system refers to potentials −3.4, 7.5, 6.0, −8 on the repeller and the three lenses, respectively. The emission current is assumed to be 500 nA

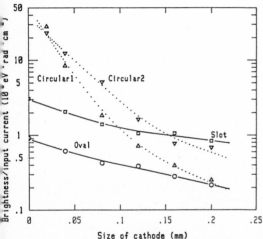

Fig. 6.18. Radiance/input current for the three emission systems as a function of the length of the squared area of the cathode. The second data set for the circular cathode system refers to potentials −3.4, 7.5, 6.0 and −8 on the repeller and the three lenses, respectively. The emission current is assumed to be 500 nA

emitting area of the cathode tip is as small as defined in Table 6.2. Such small tips may be difficult to realize and furthermore may not have sufficient long-term stability. When the emitting area is enlarged, the three emission systems react very differently. In Figs. 6.17 and 6.18 we have plotted the ratio of the radiance and brightness to the input current as a function of the cathode area (assumed to be a square) for an emission current of 500 nA. Obviously the circular emission system deteriorates rapidly with the cathode size, while the other two systems are rather robust. As a consequence, the slot system becomes the best when the linear dimension of the cathode area exceeds 0.03 mm. The circular system performs better for larger cathode areas when a negative potential is applied to the last lens aperture. The data denoted as "circular 2" in Figs. 6.17 and 6.18 are obtained with −3.4, 7.5, 6.0, −8 V on the repeller and the three lens apertures, respectively.

In view of its insensitivity to the size of the cathode, the slot lens emission system with its independent focusing in the two planes appears to be the system of choice for highest performance. The differences between the three systems in practical tests were however not so large as the calculation predicted. This is partly because a higher brightness to input current ratio can be achieved with the entrance aperture of the pre-monochromator raised to higher potentials (Sect. 5.2). With this flexibility built in, acceptable results were achieved with all three types of emission systems.

7. Lens Systems

This chapter outlines the basic concepts of the theory of electron-surface scattering insofar as they are relevant to lens design. Cylindrical deflectors focus only in the radial plane. Nonrotational symmetric lenses compensate this disadvantage. Three different lens systems are studied and their design is described in detail.

7.1 Concepts in Inelastic Electron Scattering

This chapter describes computational techniques for nonrotationally symmetric electrostatic lenses and provides a few examples of lenses which were found suitable in conjunction with the cylindrical deflectors and emission systems. We have argued earlier that such nonrotationally symmetric lenses are indispensable for attaining the optimum performance of a spectrometer incorporating energy analysers with different focusing properties in two planes. Nonrotational symmetry can be achieved in two ways. Firstly, one may use quadrupolar (or in general multipolar) lens elements, the lens segments being sections of rotationally symmetric apertures or tube elements. Different focusing in two planes is then achieved by applying different potentials to pairs of segments. Secondly, one may retain the unipotential lens elements but shape the cross-section of these in such a way that the desired focusing is achieved. The obvious advantage of the latter method is that a smaller number of independently adjustable potentials is needed. Lenses with nonrotationally symmetric apertures (such as already described for emission systems) have found little attention in the literature. The reason is presumably that, apart from the need for a three-dimensional computer simulation, they offer less flexibility with regard to the ratio of the focal length in the two planes. A specific design is therefore needed for each particular application. In the previous sections, we have studied monochromators and emission systems and have established suitable geometric parameters for these systems; we have also examined the lateral and horizontal extension of the monochromatic beam and the aperture angles of that beam when it leaves the monochromator. Thus the initial conditions for the bundle of trajectories entering the lens system are established. Furthermore we have argued in Sect. 3.3 that for the typical applications in electron energy loss spectroscopy, where the loss energy is a small fraction of the impact energy at the sample, the lens systems between the sample and the analyser and between the monochromator and the sample should consist of the same optical elements symmetrically arranged around the sample position. We now need to specify the beam parameters when it strikes the sample. The beam parameters are the impact energy, the angular apertures in the two focal planes and the lateral and vertical extension of the spot on the sample illuminated by the beam. By virtue of the symmetry of the lens system, the angular apertures of the beam impinging on the target are equal to the acceptance angles of the lens and the analysers to follow, and the spot viewed by the analyser has the same area and shape as the illuminated spot. The desired beam parameters at the sample depend in general on the nature of the scientific property which is to be investigated, and, in particular, on the physics of the inelastic scattering processes in which electrons engage at surfaces. We therefore summarise some basic concepts of inelastic scattering.

We begin with the kinematics of the scattering process. In scattering from a surface that exhibits translational symmetry along the surface plane, the parallel component of the wave vector of the electron is conserved in addition to the energy:

$$E_s = E_i + \hbar\omega \ , \tag{7.1}$$

$$k_{s\parallel} = k_{i\parallel} + Q_\parallel + G_\parallel \ . \tag{7.2}$$

Here the suffixes s and i refer to scattered and incident beams, respectively, $Q_\parallel$ is the wave vector of the elementary excitation and $\hbar\omega$ its energy. $G_\parallel$ is a vector of the surface reciprocal lattice. A positive value of $\hbar\omega$ corresponds to a process in which the electron gains energy from the surface. It is important to note that the sign of $Q_\parallel$ in (7.2) is not arbitrary. The scattering kinematics is illustrated in Fig. 7.1. The wave vector k_i of the incident electron and the surface normal

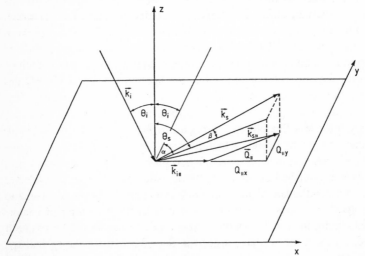

Fig. 7.1. Illustration of the scattering kinematics and the transformation of the in-plane angle α and the out-of-plane angle β into the meridional and sagittal components $Q_{\parallel x}$ and $Q_{\parallel y}$ of the surface wave vector $Q_\parallel$

define the scattering plane. The wave vector k_s of the scattered electron has a component perpendicular to the scattering plane, which is equivalent to the out-of-plane component of $Q_\parallel$ (Fig. 7.1). With the help of the scattering angles, (7.1) and (7.2) may be written as

$$Q_{\parallel x} = \frac{1}{\hbar} \sqrt{2mE_i} \left[\sqrt{1 - 2\vartheta_E} \sin(\theta_i + \alpha) - \sin\theta_i \right] , \tag{7.3}$$

$$Q_{\parallel y} = \frac{1}{\hbar} \sqrt{2mE_i(1 - 2\vartheta_E)} \sin\beta , \tag{7.4}$$

where ϑ_E is defined by

$$\vartheta_E = \hbar\omega/2E_i . \tag{7.5}$$

The angle α is the difference between θ_s and θ_i, where θ_s is defined to be the angle between the projection of k_s onto the scattering plane and the surface normal. The angle β is the angle between k_s and its projection onto the scattering plane. In electron energy loss spectroscopy, the energy $\hbar\omega$ is typically small compared with the energy of the beam E_i (i.e. $\vartheta_E \ll 1$). The wave vector of the surface excitation then depends to first order only on the scattering angles and the beam energy, not on the magnitude of the energy transfer $\hbar\omega$. A spectrum taken with fixed impact energy and angles θ_i, α, β is therefore a spectrum at a nearly constant value of $Q_\parallel$. The property of electron scattering facilitates the assignment of features in a spectrum to a specific point in the two-dimensional Brillouin-zone of the surface.

The theory of electron-solid interaction has been worked out in detail for the two most important contributions to the inelastic scattering. One contribution comes from the interaction of the electron with the potential arising from charge density fluctuations near the surface. In the following we give the scattering cross-section for this process, assuming that the displacement vector D is a linear function of the electric field $\mathcal{E}$ and that the dielectric response is a local one. In the $(Q_\parallel, \omega)$-Fourier space, linearity and locality means that one can define a dielectric tensor $\varepsilon(Q_\parallel, \omega, z)$ in terms of the Fourier components of D and $\mathcal{E}$,

$$D(Q_\parallel, z, \omega) = \varepsilon(Q_\parallel, z, \omega)\mathcal{E}(Q_\parallel, z, \omega) , \tag{7.6}$$

where z denotes the surface normal. An expression for the inelastic cross-section in a closed form was derived for the reflection of electrons from a uniform and scalar dielectric halfspace [7.1], from a halfspace with a uniaxial anisotropic dielectric function [7.2], for the transmission through a slab [7.3], and for the reflection from an isotropic dielectric halfspace covered with a uniform layer of another material [7.4; Ref. 7.5, pp. 73, 342]. The latter case is of particular interest here, as it includes the application of electron energy loss spectroscopy to inelastic scattering from adsorbed species or the surface atoms themselves. As we shall see shortly, the inelastic scattering is sharply peaked around the specular direction when $\hbar\omega \ll E_i$. In this limit the ratio of the inelastic intensity in the energy range between $\hbar\omega$ and $\hbar\omega + d\hbar\omega$ and the wave vector range between $Q_\parallel$

and $Q_{\|} + dQ_{\|}$ to the specularly reflected elastic intensity I_{el} is

$$\frac{dI_{inel}}{I_{el}} = \frac{4e^2}{\pi^2 \hbar^2} \frac{Q_{\|} v_{\perp}^2}{[v_{\perp}^2 Q_{\|}^2 + (\omega - v_{\|} \cdot Q_{\|})^2]}$$

$$\times [1 + \bar{n}(\omega)] \operatorname{Im} \frac{-1}{\tilde{\varepsilon}(Q_{\|}, \omega) + 1} d\hbar\omega \, dQ_{\|} \,. \tag{7.7}$$

Here $v_{\perp}$ and $v_{\|}$ denote the perpendicular and parallel components of the electron velocity and $\bar{n}(\omega)$ the Bose-factor. The generalized dielectric function $\tilde{\varepsilon}(Q_{\|}, \omega)$ is defined by

$$\tilde{\varepsilon}(Q_{\|}, \omega) = \varepsilon_s(\omega) \frac{1 + \Delta(\omega) \exp(-2Q_{\|}d)}{1 - \Delta(\omega) \exp(-2Q_{\|}d)} \quad \text{with} \tag{7.8}$$

$$\Delta(\omega) = \frac{\varepsilon_b(\omega) - \varepsilon_s(\omega)}{\varepsilon_b(\omega) + \varepsilon_s(\omega)} \,, \tag{7.9}$$

where $\varepsilon_b(\omega)$ and $\varepsilon_s(\omega)$ are the dielectric functions for the bulk material and for a surface layer of thickness d, respectively. In the derivation of (7.7) it is assumed that the amplitude and phase of the reflected wave at the energy $E_i - \hbar\omega$ is (nearly) the same as at E_i, which again implies that $\hbar\omega$ is small. The appropriate corrections for larger $\hbar\omega$ are given in [7.4; Ref. 7.5, pp. 73, 342]. It is obvious from the structure of (7.7) that it is applicable to all types of elementary excitations and also to arbitrary thicknesses of the surface layer. A case of special importance is that of a thin layer. A layer can be regarded as thin when

$$Q_{\|m}d \ll 1 \,, \tag{7.10}$$

where $Q_{\|m}$ is the value at the maximum of the $Q_{\|}$-dependent term in (7.7). This maximum occurs when

$$Q_{\|m} = \omega/v_{\|} \,. \tag{7.11}$$

This latter condition is the so-called surfing condition, which in other words means the electron has the maximum interaction with partial waves of a phase velocity $v_{\|}/Q_{\|}$ equal to the parallel component of the electron velocity. In the limit $Q_{\|m}d \ll 1$ the last term in (7.7) becomes

$$\operatorname{Im} \left\{ \frac{-1}{\tilde{\varepsilon}(Q_{\|}, \omega) + 1} \right\}$$

$$= Q_{\|}d \operatorname{Im} \left\{ \frac{-1}{\varepsilon_s(\omega)} \frac{\varepsilon_b^2(\omega)}{(\varepsilon_b(\omega) + 1)^2} + \varepsilon_s(\omega) \frac{1}{(\varepsilon_b(\omega) + 1)^2} \right\} \,. \tag{7.12}$$

The first term corresponds to the excitation of dipole oscillators polarized perpendicularly to the surface, the second to dipole oscillators parallel to the surface. The second term vanishes for a metal substrate where $\varepsilon_b(\omega)$ is rather large. For the typical application of electron energy loss spectroscopy, namely, to probe vibrations of adsorbed molecules on a metal substrate or substrates with a higher dielectric constant, only the first term survives or at least, it dominates. If one disregards depolarisation effects, which play a minor role except in dense layers of dipole active molecules, one may cast the remaining term into the form

$$Q_{\parallel} d \operatorname{Im} \left\{ \frac{-1}{\varepsilon_s(\omega)} \right\} = 4\pi Q_{\parallel} n_s \operatorname{Im} \{\alpha_{\perp}(\omega)\} , \tag{7.13}$$

where n_s is the surface concentration of the species and $\alpha_{\perp}(\omega)$ the polarizability perpendicular to the surface.

In practical spectroscopy one is interested in how much intensity falls into the angular aperture of a particular spectrometer. The integral over $Q_{\parallel}$ is therefore to be replaced by an integral over the angles α, β. We first transform the element $dQ_{\parallel}$ into an element of solid angle $d\alpha\, d\beta$ by means of

$$dQ_{\parallel} = k_i^2 \cos\theta_i \, d\alpha \, d\beta . \tag{7.14}$$

Inserting the equations for the scattering kinematics (7.3), (7.4) into (7.7), we obtain the following expression for the intensity scattered into a finite angular aperture

$$\frac{1}{I_{el}} \frac{dI_{inel}}{d\hbar\omega} = \frac{8(1 + \bar{n}(\omega))}{\pi a_0 E_i \cos\theta_i} n_s \operatorname{Im}\{\alpha_{\perp}(\omega)\}$$
$$\times \int d\alpha \int d\beta \frac{(\alpha \cos\theta_i + \vartheta_E \sin\theta_i)^2 + \beta^2}{(\alpha^2 + \beta^2 + \vartheta_E^2)^2} , \tag{7.15}$$

in the limit $\vartheta_E \ll 1$, where a_0 is the Bohr radius.

High scattering cross sections are obtained near grazing incidence. The extent of the angular distribution then becomes unequal along α and β, with the angular spread being broader in β. We also note that the maximum of the angular distribution is near the specular direction ($\alpha = 0$) but slightly shifted towards smaller angles ($\alpha < 0$). The form of the integral in (7.15) with the rectangular-shaped angular space is particular suitable for the rectangular-shaped aperture in spectrometers featuring cylindrical deflectors. Typical values of the integral when the integration is performed in symmetric intervals $-\alpha_c < \alpha < \alpha_c$ and $-\beta_c < \beta < \beta_c$ around the specularly reflected beam are tabulated in Table 7.1. The integral can be expressed in closed form for a circular angular aperture centred around the specular beam. After introducing the polar angle ϑ and the azimuthal angle φ by the transformation

$$\alpha = \vartheta \cos\varphi , \quad \beta = \vartheta \sin\varphi , \tag{7.16}$$

and integrating over the azimuthal angle φ and the polar angle ϑ up to a maximum angle ϑ_c, one obtains, for the inelastic intensity

Table 7.1. The integral in (7.15)

$$\frac{2}{\pi} \int_{-\hat{\alpha}_c}^{\hat{\alpha}_c} d\hat{\alpha} \int_{-\hat{\beta}_c}^{\hat{\beta}_c} d\hat{\beta} \frac{(\hat{\alpha} \cos \theta_i + \sin \theta_i)^2 + \hat{\beta}^2}{(1 + \hat{\alpha}^2 + \hat{\beta}^2)^2}$$

for a set of characteristic parameters $\hat{\alpha}_c$, $\hat{\beta}_c$ and θ_i. The reduced units $\hat{\alpha}$, $\hat{\beta}$ are defined as $\hat{\alpha} = \hat{\alpha}/\vartheta_E$, $\hat{\beta} = \hat{\beta}/\vartheta_E$ with $\vartheta_E = \hbar\omega/2E_i$

	$\theta = 50°$			$\theta = 60°$			$\theta = 70°$	
$\hat{\alpha}_c$	$\hat{\beta}_c$	Integral	$\hat{\alpha}_c$	$\hat{\beta}_c$	Integral	$\hat{\alpha}_c$	$\hat{\beta}_c$	Integral
0.2	0.2	0.5797	0.2	0.2	0.734	0.2	0.2	0.08598
0.2	0.4	0.1145	0.2	0.4	0.1432	0.2	0.4	0.1667
0.2	0.6	0.1677	0.2	0.6	0.2067	0.2	0.6	0.2385
0.2	0.8	0.2161	0.2	0.8	0.2623	0.2	0.8	0.3
0.2	1	0.2587	0.2	1	0.3099	0.2	1	0.3517
0.4	0.2	0.1107	0.4	0.2	0.1384	0.4	0.2	0.161
0.4	0.4	0.2188	0.4	0.4	0.2705	0.4	0.4	0.3127
0.4	0.6	0.321	0.4	0.6	0.3914	0.4	0.6	0.4487
0.4	0.8	0.4144	0.4	0.8	0.498	0.4	0.8	0.5662
0.4	1	0.4972	0.4	1	0.59	0.4	1	0.6657
0.6	0.2	0.1553	0.6	0.2	0.1908	0.6	0.2	0.2198
0.6	0.4	0.3072	0.6	0.4	0.3737	0.6	0.4	0.4279
0.6	0.6	0.4516	0.6	0.6	0.5423	0.6	0.6	0.6162
0.6	0.8	0.5844	0.6	0.8	0.6925	0.6	0.8	0.7806
0.6	1	0.7029	0.6	1	0.8231	0.6	1	0.9211
0.8	0.2	0.1913	0.8	0.2	0.2306	0.8	0.2	0.2627
0.8	0.4	0.3789	0.8	0.4	0.4527	0.8	0.4	0.5128
0.8	0.6	0.5581	0.8	0.6	0.6589	0.8	0.6	0.741
0.8	0.8	0.7239	0.8	0.8	0.8443	0.8	0.8	0.9425
0.8	1	0.8731	0.8	1	1.007	0.8	1	1.117
1	0.2	0.2198	1	0.2	0.2602	1	0.2	0.2932
1	0.4	0.4359	1	0.4	0.5117	1	0.4	0.5735
1	0.6	0.6431	1	0.6	0.7467	1	0.6	0.8311
1	0.8	0.836	1	0.8	0.9599	1	0.8	1.061
1	1	1.011	1	1	1.149	1	1	1.261
2	2	2.322	2	2	2.438	2	2	2.531
2	4	3.188	2	4	3.303	2	4	3.396
2	6	3.558	2	6	3.672	2	6	3.765
2	8	3.756	2	8	3.87	2	8	3.963
2	10	3.88	2	10	3.993	2	10	4.086
4	2	2.808	4	2	2.816	4	2	2.823
4	4	4.061	4	4	4.033	4	4	4.009
4	6	4.68	4	6	4.64	4	6	4.607
4	8	5.038	4	8	4.994	4	8	4.957
4	10	5.269	4	10	5.223	4	10	5.184
6	2	2.983	6	2	2.935	6	2	2.897
6	4	4.405	6	4	4.29	6	4	4.195
6	6	5.169	6	6	5.027	6	6	4.909
6	8	5.638	6	8	5.485	6	8	5.357
6	10	5.953	6	10	5.794	6	10	5.663

Table 7.1. (cont.)

$\hat{\alpha}_c$	$\theta = 50°$ $\hat{\beta}_c$	Integral	$\hat{\alpha}_c$	$\theta = 60°$ $\hat{\beta}_c$	Integral	$\hat{\alpha}_c$	$\theta = 70°$ $\hat{\beta}_c$	Integral
8	2	3.072	8	2	2.992	8	2	2.929
8	4	4.583	8	4	4.415	8	4	4.276
8	6	5.432	8	6	5.223	8	6	5.049
8	8	5.976	8	8	5.748	8	8	5.559
8	10	6.354	8	10	6.117	8	10	5.919
10	2	3.126	10	2	3.026	10	2	2.948
10	4	4.69	10	4	4.488	10	4	4.32
10	6	5.594	10	6	5.34	10	6	5.127
10	8	6.19	10	8	5.909	10	8	5.674
10	10	6.615	10	10	6.319	10	10	6.072

$$\frac{1}{I_{el}} \frac{dI_{inel}}{d\hbar\omega} = \frac{4(1 + \bar{n}(\omega))}{a_0 E_i \cos\theta_i} n_s \, \text{Im}\{\alpha_\perp(\omega)\}$$
$$\times \left[(\sin^2\theta_i - 2\cos^2\theta_i) \frac{\vartheta_c^2}{\vartheta_c^2 + \vartheta_E^2} + (1 + \cos^2\theta_i) \ln\left(1 + \frac{\vartheta_c^2}{\vartheta_E^2}\right) \right] . \tag{7.17}$$

For well-designed lens systems and typical energy losses, ϑ_c can be greater than ϑ_E, where the ϑ_c-dependent term on the right-hand side of (7.17) is nearly saturated, save for the logarithmic term. For some commercially available spectrometers, with a rather large scattering chamber and consequently a smaller ϑ_c, the limit $\vartheta_c \ll \vartheta_E$ is of interest. The integral in (7.15) then becomes proportional to ϑ_c^2 and the inelastic intensity in the low temperature limit ($\bar{n}(\omega) = 0$) is given by

$$\frac{1}{I_{el}} \frac{dI_{inel}}{d\hbar\omega} = \frac{32 E_i}{a_0 (\hbar\omega)^2} \frac{\sin^2\theta_i}{\cos\theta_i} \vartheta_c^2 n_s \, \text{Im}\{\alpha_\perp(\omega)\} , \quad \vartheta_c \ll \vartheta_E . \tag{7.18}$$

It is interesting to compare this result with infrared reflection absorption spectroscopy on a perfect metal surface. There the change in the reflectivity for p-polarized light ΔR_p on a perfect metal surface $[\varepsilon_b(\omega) = -\infty]$ is given by

$$\Delta R_p = \frac{8\omega}{c} \frac{\sin^2\theta_i}{\cos\theta_i} n_s \, \text{Im}\{\alpha_\perp(\omega)\} . \tag{7.19}$$

The different scaling of the prefactors in ΔR_p and I_{inel} results in a substantially different sensitivity between infrared reflection-absorption spectroscopy and electron energy loss spectroscopy along the frequency scale. While the optical detection of vibrational surface modes in the far-infrared is nearly impossible, these modes become intense features in an energy loss spectrum. With this remark we conclude our discussion of some basic features of dipole scattering and turn to the second important contribution to the inelastic intensity for which the term *impact scattering* has become common.

Inelastic impact scattering is a direct consequence of the atomic nature of matter and the fact that the atoms move about their equilibrium position by virtue of their thermal energy. As the motion of the atoms is slow compared to the transient time of an electron on each atom, the electron is diffracted from the ensemble of atoms at their instantaneous positions $R_n(t)$. When these positions are expanded into phonon creation and annihilation operators, the cross-section for phonon creation and annihilation for a primitive lattice in the Born approximation [7.6] may be written

$$\frac{d\sigma}{d\hbar\omega\, d\Omega} = \left(\frac{m}{h}\right)^2 |f(\mathbf{k}_s - \mathbf{k}_i)|^2 \frac{N_s}{2M\hbar\omega}$$
$$\times \left| \sum_{Q_\parallel, s, l} (\mathbf{k}_s - \mathbf{k}_i) \cdot e_s(Q_\parallel, z_l) \exp[iz_l(k_{sz} + k_{iz})] \right|^2$$
$$\times [(\bar{n}(\omega) + 1)\delta(\hbar\omega_s(Q_\parallel) - \hbar\omega)\delta_{k_{s\parallel} - k_{i\parallel} + Q_\parallel, G_\parallel}$$
$$+ \bar{n}(\omega)\delta(\hbar\omega_s(Q_\parallel) + \hbar\omega)\delta_{k_{s\parallel} - k_{i\parallel} - Q_\parallel, G_\parallel}] \,. \tag{7.20}$$

Here $f(\mathbf{k}_s - \mathbf{k}_i)$ is the elastic scattering amplitude and $e_s(Q_\parallel, z_l)$ is the phonon eigenvector of the phonon branch s in the layer l where z_l denotes the position of that layer with respect to the surface. N_s is the number of surface unit cells and M the mass of the atoms. The last terms in (7.20) ensure momentum conservation in energy loss and gain processes, respectively. Because of the strong multiple scattering processes, the simple expression (7.20) for the phonon cross-section is of no quantitative value. Several properties of (7.20) are however pertinent to a fully dynamical treatment of the scattering process [7.7]. In particular, phonons can contribute only when their amplitude has a component within the scattering plane spanned by $\mathbf{k}_s$ and $\mathbf{k}_i$. If this scattering plane is aligned with a mirror plane of the crystal, phonon modes that are odd with respect to that mirror plane do not contribute to the cross-section. The same is true of course for localised vibrational modes. This selection rule is independent of the Born approximation made in (7.20) and can be deduced quite generally from the time inversion symmetry of the problem [Ref. 7.5, pp. 117 ff.]. Equation (7.20) also shows that the relative intensity of one-phonon processes compared to the elastic scattering increases as k^2, i.e., proportional to the impact energy. High cross-sections for phonon scattering are therefore frequently found at higher impact energies, unlike dipole losses, the intensity of which is inversely proportional to the impact energy. When the energy becomes too large, however, multiple phonon events scaling as E^n with $n \geq 2$ contribute a larger background to the spectrum. The useful regime of impact energies for phonon spectroscopy has therefore an upper boundary at about 200 eV.

A final consequence of (7.20) is that, because the exchange of perpendicular momentum with the surface is greater than that of the parallel component of the momentum, vertically polarized modes have a larger cross-section than parallel modes. This statement does however, need further qualification. In reality, the Born approximation gives an inadequate account of the scattering process. Strong, primarily intralayer, multiple scattering makes the cross-section for a particular

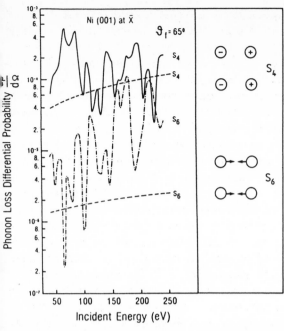

Fig. 7.2. Dependence of the cross-section for inelastic (impact) scattering from two different phonon modes. The inserts show the polarisation of the modes in the first atom layer. The Born approximation (*dashed lines*) predicts a higher cross-section for the vertically polarised mode through the entire energy range. In reality, multiple scattering events cause the cross-section to oscillate around the value calculated in the Born approximation [7.10]

mode oscillate rather sharply as a function of the impact energy (Fig. 7.2). These oscillations are the typical consequence of multiple interference processes. While the Born approximation is a reasonable description of the average dependence of the cross-section on the energy and polarization of the mode, the details are more complex. In particular, one may encounter situations where the intensity of a parallel mode exceeds that of a perpendicular mode. Such situations are in fact in practice even more frequent than in the particular example shown in Fig. 7.2. From this, we conclude that the sharp oscillations in the cross-sections for a particular phonon mode require a spectrometer in which the impact energy can easily be varied over a wide energy range.

7.2 Image Formation and Momentum Resolution

In the previous section we have considered scattering processes on well-ordered surfaces. In such cases, one typically probes the elementary excitations as a function of their momentum parallel to the surface and one desires a momentum resolution of a few percent of the surface Brillouin-zone. Quite frequently

– in fact in many practical applications – the surfaces are disordered. As a consequence, $Q_\parallel$ is no longer a good quantum number. Sharp spectral features of considerable dispersion on the ordered surface become broad bands when the surface is disordered. The widths of the bands represent the amount of dispersion in the ordered case. On the other hand, dispersionless excitations, which are dispersionless because they are localised in space, remain narrow lines in the spectra of the disordered surfaces. In both cases the same spectrum is observed at different scattering angles and the intensity of the inelastic signal is proportional to the solid angle accepted by the analyser. We shall see later that even when ordered surfaces are probed, the desire for sufficient momentum resolution is rarely a point of serious concern in lens design. In other words, most lenses will provide acceptance angles that are smaller rather than larger than the maximum solid angle derived from consideration of reasonable momentum resolution. For most applications, therfore, we attempt to maximise the acceptance angle at the target. We recall that, for the majority of applications, the lenses between the monochromator and sample and between the sample and the analyser should be built symmetrically around the sample (Sect. 3.3). The convergence angles α and β of the beam impinging on the sample are hence equal to those angles of the subsequent lens system and analyser within which scattered electrons are accepted to eventually pass through the exit slit of the analyser and enter the detector. In the following dicscussion, therefore, we need discuss only one lens system, rather than two, and we identify the convergence angles α_i and β_i of the incoming beam at the sample with the acceptance angles α_s and β_s of the lens system that collects the scattered electrons. As one wants to maximise the acceptance angles and thus the angles of the incoming beam, one needs to process the beam in such a way that the lateral extension of the beam at the sample is small. This simply follows from phase space conservation (3.20). The smallest cross-section of the beam at the sample is achieved when the trajectories are such that an image of a real aperture is formed at the sample. Remembering that cylindrical deflectors form an image of the entrance slit at the exit slit position only in the radial ("horizontal") plane, the maximum acceptance angle is achieved by forming an image of the *exit* slit of the monochromator in the *horizontal* plane and simultaneously an image of the *entrance* slit of the monochromator in the vertical plane, so that one has a sharp image of the entrance slit at the sample (Fig. 7.3) [7.8]. Obviously this task cannot be performed by lenses circularly symmetric around the optic axis. The design of such lenses therefore cannot build upon the established experience in the optics of circularly symmetric lenses. Instead, we need fully three-dimensional computer simulations and efficient codes for that purpose [7.9].

It should be mentioned that maximisation of the acceptance angle, and thus the need for a sharp intermediate image at the sample, is not identical with maximisation of the total monochromatic current at the detector, when the analyser is viewing the primary beam directly. In this latter case, the lens systems can be operated in such a way as to have their foci at infinity, so that a parallel beam is formed at the sample. A high monochromatic current at the detector and a

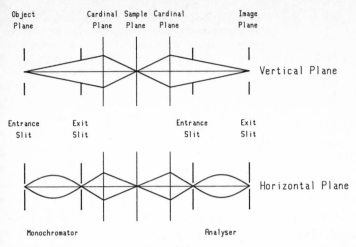

Fig. 7.3. Schematic illustration of the beam guidance in an electron energy loss spectrometer employing cylindrical deflectors as energy dispersive elements [7.8]. In reality the trajectories are not straight lines in the lens and sample area because of the continuously varying potential. Focusing as shown here provides for the maximum acceptance angle at the sample while maintaining high transmission

small loss of current between the sample and the detector position characterises the quality of a lens system, though insufficiently, since it is the product of the acceptance angle with the detector current which is the significant figure of merit.

Having argued in favour of a large acceptance angle, we must now specify more explicitly how large an acceptance angle can be tolerated, or how large a $Q_\parallel$-space should be sampled in order to have both a high signal and small errors in the values of the characteristic energy losses. Obviously the error will be large where the dispersion of a characteristic loss is large. A particularly large error of this type arises when the dispersion of an acoustic surface phonon is probed. There, the frequency depends qualitatively on $Q_\parallel$ as

$$\omega(Q_\parallel) = \sin(Q_\parallel \tfrac{\pi}{2}) \ . \tag{7.21}$$

For simplicity we have assumed that the dispersion is isotropic. The wave vector $Q_\parallel$ and the frequency are in reduced units so that the zone boundary is at $Q_\parallel = 1$ and the frequency there is $\omega = \sin(\pi/2) = 1$. Suppose one is now sampling a finite space in the $Q_\parallel$-plane, which is rectangular in shape with sides $\Delta Q_{\parallel x}$ and $\Delta Q_{\parallel y}$. Such a rectangularly shaped $Q_\parallel$-space is actually probed by the spectrometer since the kinematics of the scattering (7.3) and (7.4) transform the acceptance α_i and β_i into a rectangular $Q_\parallel$-space via

$$\Delta Q_{\parallel x} = \frac{1}{\hbar} \sqrt{2mE_0(1 - 2\vartheta_E)} \, \cos(\theta_i)\alpha_i \quad (\alpha_i \ll 1) \ , \tag{7.22}$$

$$\Delta Q_{\parallel y} = \frac{1}{\hbar} \sqrt{2mE_0(1 - 2\vartheta_E)} \, \beta_i \quad (\beta_i \ll 1) \ , \tag{7.23}$$

125

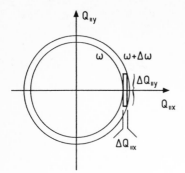

Fig. 7.4. This figure illustrates the frequency error introduced by measuring the dispersion of a characteristic loss where the frequency depends on the modulus of $Q_{\parallel}$, due to the sampling of a finite $Q_{\parallel}$-area spanned by the rectangle $\Delta Q_{\parallel x}$ and $\Delta Q_{\parallel y}$. An optimum $Q_{\parallel}$-area is obtained with $\Delta Q_{\parallel y} > \Delta Q_{\parallel x}$

with ϑ_E defined as before. For excitation of a mode with a dispersion as in (7.21), the spectrometer measures an average frequency of

$$\omega(Q_{\parallel}) = \int_{-\Delta Q_{\parallel x}/2}^{+\Delta Q_{\parallel x}/2} dQ'_{\parallel x} \int_{-\Delta Q_{\parallel y}/2}^{\Delta Q_{\parallel y}/2} dQ'_{\parallel y} \sin[(Q_{\parallel} + Q'_{\parallel})\tfrac{\pi}{2}] . \tag{7.24}$$

It is clear from Fig. 7.4 that one can tolerate much larger $\Delta Q_{\parallel y}$ than $\Delta Q_{\parallel x}$. This is in accord with the optical properties of spectrometers with cylindrical deflectors and the focusing discussed earlier. Table 7.2 shows a set of values for the integral (7.24) for $\Delta Q_{\parallel x} = 0.005; 0.01; 0.015$ and $\Delta Q_{\parallel y} = 0.05; 0.1; 0.15$ and the error compared to the true value of the frequency. Obviously the error is largest for small $Q_{\parallel}$. In typical applications, one begins to resolve phonon features in a spectrum for $Q_{\parallel} \geq 0.2$. Even then the error is smaller than 10% when $\Delta Q_{\parallel y}$ is 0.15, for which one samples as much as 15% of the Brillouin zone in the y-direction, and $\Delta Q_{\parallel y} = 0.1$ causes only 4% error at the most. In the x-direction too, $\Delta Q_{\parallel x}$ can be made quite large without introducing excessive errors. We have restricted Table 7.2 to $\Delta Q_{\parallel x} \leq 0.015$ because it is difficult to achieve large enough acceptance angles α_i, while on the other hand the acceptance angle β_i can be much larger. This simply follows from the fact that the ratio of distance of the object to the cardinal plane to the distance of the image to the cardinal plane is much larger in the vertical plane than in the horizontal (Fig. 7.13).

When the lenses form images as shown in Fig. 7.3, the monochromator and the analyser are symmetrically built, and space charge distortions are disregarded, then the current at the detector can be almost as high as the current at the sample position when the sample is removed and the analyser views directly the monochromatic beam ("direct beam position"). If the analyser is adjusted to have the same energy resolution (see also Sect. 3.2) as the monochromator and the transmitted energy distribution resembles a gaussian, one may estimate the ratio of the current at the detector I_D to the current at the sample position I_s to be

$$\frac{I_D}{I_s} \approx \frac{1}{\sqrt{1 + (\Delta E_M / \Delta E_A)^2}} \frac{\int_{-1}^{1}(1 - \alpha^2)^2 d\alpha}{\int_{-1}^{1}(1 - \alpha^2)d\alpha} = \frac{1}{2}\sqrt{\frac{2}{7}} \approx 27\% . \tag{7.25}$$

Table 7.2. Mean measured frequency and the derivation from the true value of the frequency when the frequency is measured with a spectrometer sampling a finite, rectangularly shaped area in $Q_\|$ of dimensions $\Delta Q_{\|x}$, $\Delta Q_{\|y}$ when the frequency has a dispersion $\omega(Q_\|) = \sin(Q_\| \pi/2)$. Frequency and wave vector are in reduced units

$\Delta Q_{\|x} = 0.005, \Delta Q_{\|y} = 0.05$			$\Delta Q_{\|x} = 0.005, \Delta Q_{\|y} = 0.1$			$\Delta Q_{\|x} = 0.005, \Delta Q_{\|y} = 0.15$		
Q	Frequency	Error	Q	Frequency	Error	Q	Frequency	Error
0	0.0401	0.0401	0	0.0795	0.0795	0	0.119	0.119
0.1	0.164	0.00716	0.1	0.18	0.024	0.1	0.204	0.0476
0.2	0.314	0.00469	0.2	0.323	0.0137	0.2	0.337	0.0277
0.3	0.458	0.00423	0.3	0.464	0.1	0.3	0.473	0.0193
0.4	0.592	0.00426	0.4	0.596	0.00822	0.4	0.602	0.0147
0.5	0.712	0.0445	0.5	0.714	0.00723	0.5	0.719	0.0118
0.6	0.814	0.00466	0.6	0.816	0.00659	0.6	0.819	0.00974
0.7	0.896	0.00484	0.7	0.897	0.00612	0.7	0.899	0.00821
0.8	0.956	0.00497	0.8	0.957	0.00572	0.8	0.958	0.00696
0.9	0.993	0.00501	0.9	0.993	0.00534	0.9	0.994	0.00589
1	1	0.00495	1	1	0.00494	1	1	0.00492

$\Delta Q_{\|x} = 0.01, \Delta Q_{\|y} = 0.05$			$\Delta Q_{\|x} = 0.01, \Delta Q_{\|y} = 0.1$			$\Delta Q_{\|x} = 0.01, \Delta Q_{\|y} = 0.15$		
Q	Frequency	Error	Q	Frequency	Error	Q	Frequency	Error
0	0.0413	0.0413	0	0.0802	0.0802	0	0.119	0.119
0.1	0.164	0.00735	0.1	0.181	0.0242	0.1	0.0204	0.0477
0.2	0.314	0.00478	0.2	0.323	0.0138	0.2	0.337	0.0278
0.3	0.458	0.00429	0.3	0.0464	0.0101	0.3	0.473	0.0193
0.4	0.592	0.0043	0.4	0.596	0.00826	0.4	0.602	0.147
0.5	0.712	0.00447	0.5	0.714	0.00725	0.5	0.719	0.0118
0.6	0.814	0.00468	0.6	0.816	0.00661	0.6	0.819	0.00976
0.7	0.896	0.00486	0.7	0.897	0.00613	0.7	0.899	0.00822
0.8	0.956	0.00497	0.8	0.957	0.00573	0.8	0.958	0.00697
0.9	0.993	0.00501	0.9	0.993	0.00535	0.9	0.994	0.00589
1	1	0.00495	1	1	0.00494	1	1	0.00492

$\Delta Q_{\|x} = 0.015, \Delta Q_{\|y} = 0.05$			$\Delta Q_{\|x} = 0.015, \Delta Q_{\|y} = 0.1$			$\Delta Q_{\|x} = 0.015, \Delta Q_{\|y} = 0.15$		
Q	Frequency	Error	Q	Frequency	Error	Q	Frequency	Error
0	0.0429	0.0429	0	0.0812	0.0812	0	0.12	0.12
0.1	0.164	0.00766	0.1	0.181	0.0245	0.1	0.204	0.048
0.2	0.314	0.00493	0.2	0.323	0.0139	0.2	0.337	0.028
0.3	0.458	0.00438	0.3	0.464	0.0102	0.3	0.473	0.0194
0.4	0.592	0.00437	0.4	0.596	0.00833	0.4	0.603	0.148
0.5	0.712	0.00452	0.5	0.714	0.0073	0.5	0.719	0.0118
0.6	0.814	0.00471	0.6	0.816	0.00664	0.6	0.819	0.00979
0.7	0.896	0.00488	0.7	0.897	0.00616	0.7	0.899	0.00824
0.8	0.956	0.00499	0.8	0.957	0.00575	0.8	0.958	0.00698
0.9	0.993	0.00501	0.9	0.993	0.00535	0.9	0.994	0.0059
1	1	0.00495	1	1	0.00494	1	1	0.00492

The first term is the contribution from the energy transmission function (3.48) subject to the condition (3.53) for an optimum match of monochromator and analyser. The second term arises from the transmission function as a function of the angle $\hat{\alpha} = \alpha/\sqrt{3/4r_0}$ for an ideal cylindrical deflector (3.29) and for a pair of two sequentially arranged deflectors. In writing (7.25) we have assumed that the transmission function with respect to α of two sequential deflectors is the square of the transmission function of one, which is the case when the angle vs position correlation is lost between the deflectors (Sects. 3.4 and 3.5). The ratio of 27% is an upper limit which requires a perfect match between the image of the monochromator exit slit and the analyser entrance slit. The best experimental values of I_D/I_s are about 30%, in agreement with the estimate. A reduction of the performance of the system occurs if the cross-section of the bundle of trajectories arriving at the entrance slit of the analyser exceeds the size of the slit, even when the first-order image of the exit slit of the monochromator fits the entrance slit. Furthermore the angular distribution of the electrons entering the analyser may extend to angles that are not transmitted by the analyser. An assessment of the quality of a lens system must therefore take into account the higher-order aberration terms in the image formation, in addition to the first-order properties. In the following we therefore study the nature of the aberration terms of nonspherically symmetric lenses.

In the mathematical description of the process of image formation one typically relates the cartesian coordiantes x', y' of a trajectory in the first-order image plane perpendicular to the optic z-axis to the cartesian coordinates x, y of the object and the coordinates in the pupil. Since no well-defined pupil exists in electron spectrometers (Sect. 7.4), the coordinates in the pupil are replaced by the previously defined angles α and β, which are the angles of the trajectories in the object plane projected into the xy- and yz-plane, respectively. The image plane is defined as the plane where the first-order terms of α and β vanish in an expansion of the image-object function as a polynomial

$$x' = f_x(x, y, \alpha, \beta) = \mathcal{O}(x, y, \alpha, \beta) + \mathcal{O}^2(x, y, \alpha, \beta) \dots , \tag{7.26}$$

$$y' = f_y(x, y, \alpha, \beta) = \mathcal{O}(x, y, \alpha, \beta) + \mathcal{O}^2(x, y, \alpha, \beta) \dots . \tag{7.27}$$

In writing out the most general form of the expansion one has 9 second-order and 20 third-order terms for each of the two equations. For a lens system with C_{2v} symmetry with both the xz- and the yz-planes being mirror planes, all 9 second-order terms and 12 of the third-order terms vanish. The remaining terms of the image equations are, up to third order:

$$x' = - [M_x x + C_{\alpha\alpha\alpha}\alpha^3 + C_{\alpha\alpha x}\alpha^2 x + C_{\alpha xx}\alpha x^2 + C_{xxx}x^3$$
$$+ C_{\alpha\beta\beta}\alpha\beta^2 + C_{\beta\beta x}\beta^2 x + C_{\alpha yy}\alpha y^2 + C_{xyy}xy^2] , \tag{7.28}$$

$$y' = - [M_y y + C_{\beta\beta\beta}\beta^3 + C_{\beta\beta y}\beta^2 y + C_{\beta yy}\beta y^2 + C_{yyy}y^3$$
$$+ C_{\beta\alpha\alpha}\beta\alpha^2 + C_{\alpha\alpha y}\alpha^2 y + C_{\beta xx}\beta x^2 + C_{yxx}yx^2] . \tag{7.29}$$

The sign of the coefficients are chosen so that the key parameters and the magnification are positive quantities. For a circularly symmetric lens the corresponding third-order terms in (7.28) and (7.29) become equal, the terms C_{xyy}, C_{yxx} and $C_{\alpha\beta\beta}$ vanish, and the coma terms $C_{\alpha\alpha x}$, $C_{\beta\beta x}$ are related so that one has five independent third-order aberrations. These are the spherical aberrations (here $C_{\alpha\alpha\alpha}$, $C_{\beta\beta\beta}$), the coma (here $C_{\alpha\alpha x}$, $C_{\beta\beta y}$), the tangential or meridional astigmatism (here $C_{\alpha xx}$, $C_{\beta yy}$), the sagittal astigmatism or sagittal field curvature (here $C_{\alpha yy}$, $C_{\beta xx}$), and the distortion (here C_{xxx}, C_{yyy}) [7.11]. The geometrical meaning of these third-order aberrations have been illustrated earlier in Fig. 3.2 except for the sagittal astigmatism.

As long as one is not interested in spatial resolution, the distortions of the image at the sample are not very important. The question of prime interest here is how the third-order aberrations affect the transmission of the spectrometer. We discuss this issue with the help of Fig. 7.5. There, the object plane, the plane of the intermediate image at the sample and the image plane are drawn. The image plane could for example be the exit slit of the analyser, when the plane of the drawing is the vertical plane of the spectrometer. The object plane is then in the *entrance* slit of the monochromator. The effect of third-order aberrations on the transmission is illustrated by considering the angular aberration term $-C_s\beta^3$ (with $C_s = C_{\beta\beta\beta}$). The lenses are here assumed to be thin with no loss of generality in the considerations that follow. The beam emerging from the object point on the optic axis at an angle β with the optic axis (drawn as a full line) arrives in the intermediate image plane at a position $y' = -C_s\beta^3$. Because of the time reversal symmetry, an electron could also emerge from the intermediate image plane and travel backwards along the same path in the reverse direction. We now assume that the second lens on the right hand side of Fig. 7.5 is symmetric relative to the sample so that the sample plane is a mirror plane (σ_p). The plane orthogonal to the plane of the drawing along the optic axis is likewise a mirror plane (σ_y). By applying time inversion, the σ_p-operation, and the σ_y-operation we generate the trajectory shown as the dash-dotted line in Fig. 7.5 as a possible correct

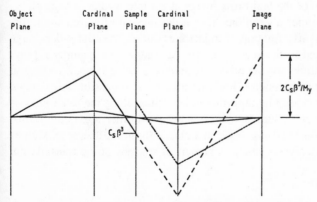

Fig. 7.5. Propagation of third-order image aberrations in a double lens system which is symmetric around the intermediate image plane ("sample plane")

trajectory. The trajectory begins in the intermediate image plane at $y = C_s\beta^3$ and ends in the second image plane on the optic axis. This latter trajectory starts at a point displaced by the distance $2C_s\beta^3$ from the position where the original trajectory meets the intermediate image. Consequently the first order image of the original trajectory in the second image plane is displaced from the optic axis by the amount $2C_s\beta^3/M_y$, when M_y is the linear magnification of the first lens and thus M_y^{-1} the magnification of the second. We see from these considerations that the third-order aberrations in the first imaging process lead to aberrations in the second image, which are twice as large even when the second introduces no further aberration. It is also clear from Fig. 7.5 that, because of the shift involved in the generation of the dotted line as the continuing trajectory of the original trajectory (full line) from the correct time inversed and σ_y, σ_p-generated trajectory we should also take coma, astigmatism and distortion errors of the second lens system into account. Since these terms include the y-coordinate at the intermediate image, these contributions would be of higher order though and are disregarded here. To lowest order in the aberrations the y''-position in the image plane is

$$y'' = 2C_s\beta^3/M_y . \tag{7.30}$$

In order to relate this aberration at the second image to the transmission, the ratio of y'' to the size of the first-order image

$$\hat{C}_s'' = \frac{y''}{h/2} = \frac{2C_s\beta^3}{M_y}\frac{1}{h/2} \tag{7.31}$$

is a useful quantity. One may also define the corresponding ratio for the intermediate image, which is

$$\hat{C}_s' = \frac{C_s\beta^3}{M_y h/2} , \tag{7.32}$$

since $M_y h/2$ is the size of the first-order image at the intermediate image plane. We thus see that third-order aberrations have a detrimental effect on the transmission, and that the relative fraction of trajectories outside the first-order image is approximately doubled in the second image. As long as the quantity C_s' and the corresponding quantities for the other image aberrations are small compared with unity, the overall transmission is high. In principle it is possible to calculate the loss in transmission caused by third-order aberrations along the same lines as for the cylindrical deflector. Because of the many third-order aberration terms, the integration limits cannot be expressed in a closed form. Furthermore, higher-order aberrations are not always small. A numerical analysis of the transmission is therefore indicated.

7.3 Examples of Lens Systems

In this section we present results for three different lens systems which were selected from a larger series of about 100 different lens systems because their properties were found to be particularly suitable for use in connection with electron spectrometers. The technical part of the computer simulation has already been described for the cathode systems (6.2). An additional point of concern here is the appropriate choice of the size of the mesh. A sensitive test is to compare the potentials that create the image at the desired image position, the first-order properties, and the principal aberration terms for two different mesh sizes. The standard mesh size for the lens simulation was $\Delta x = \Delta y = \Delta z = 1$ mm. In Table 7.3 the key parameters are compared with those given by a second calculation of the same lens with a mesh size of $\Delta x = \Delta y = \Delta z = 0.5$ mm. In both calculations the same number of integration steps was used in the trajectory calculations (~ 2000). We see that the results are nearly identical. Since the calculations with the smaller mesh occupied 8 times as much memory space for the program solving the Laplace equation, and also required at least an order of magnitude more computing time, all subsequent calculations were performed with the 1 mm mesh.

Table 7.3. Comparison of results for a lens system for two different mesh sizes, $\Delta x = \Delta y = \Delta z = 1$ mm and $\Delta x = \Delta y = \Delta z = 0.5$ mm. The results refer to lens I (Fig. 7.6)

Δx [mm]	V_1 [eV]	V_2 [eV]	V_3 [eV]	M_x	M_y
1	+0.19	1.56	5.3	−0.528	−0.108
0.5	+0.195	1.53	5.3	−0.519	−0.107

Δx [mm]	C_{sx} [mm/rad^3]	C_{sy} [mm/rad^3]	C_{cy} [rad^{-2}]	C_{ay} [mm^{-1}rad^{-1}]
1	98.4	12100	223	1.33
0.5	82.2	11500	218	1.17

All the lens systems to be discussed in the following are three-element lenses with the last lens element at the same potential as the target. The potential of the last element must be equal to that of the target in order to have a field-free region around the sample, so that one has well-defined scattering kinematics. For the same reason, the last lens is always thick so that there is no field penetration from the other lens elements (Fig. 7.6); any such penetration would lead to bending of the trajectories near the sample and thus again to ill-defined scattering kinematics. In the simulation the target is completely enclosed by a target chamber held at the target potential. The potentials of the two additional lens elements labelled 1 and 2 in Fig. 7.6 can be chosen independently. This allows us to adjust the foci in the horizontal and vertical planes independently. The range in which the ratio of the focal lengths in the vertical and the horizontal plane can be varied

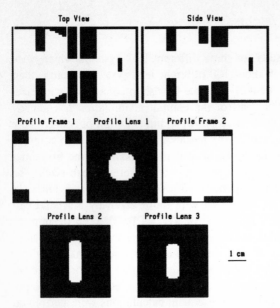

Fig. 7.6. Cross-section of lens I as used in this computer simulation. Only the inner profiles of the lens and the frame simulate the real lens system correctly. In reality the target chamber has a different shape which should have no effect on the trajectories since the target chamber is a space enclosed by metal plates of the same potential. The frames between lens element 1 and the exit slit of the monochromator on one side and element 2 on the other are electrically connected to element 1 and serve in practice to screen the insulators needed for mounting the lens

depends on the shape of the profiles of lens elements 1–3. We have argued earlier that one wishes to create an image of the exit slit of the monochromator at the target in the horizontal plane, and an image of the entrance slit of the (first) monochromator in the vertical. If this was exactly true, one would need not much flexibility in the variability of the ratio of the focal lengths of the lens in the two planes. We recall, however, that the particular focusing described above would be appropriate only in the absence of space charge. We have seen in Sect. 5.1 that the space charge spreads the beam in the vertical plane and thus may be partly compensated by applying correction potentials to the cover plates. In order to retain the possibility of adjusting the focal length of the lens system so that a space charge optimised beam emerges from the monochromator, flexibility in the vertical focal length is needed. The different profiles of lens element 1 on the one hand and of the elements 2 and 3 on the other secure this flexibility. If for example the potential on element 1 is adjusted so that it is approximately at the potential of the space occupied by element 1 in the absence of that element, then only the combination of lenses 2 and 3 serves for focusing. The large aspect ratio of height to width of the latter two elements then gives a lens system with very little vertical focusing. On the other hand, when lens element 2 is held at the potential of the corresponding space in the absence of that element, one has strong vertical focusing because of the circularly shaped lens element 1. Thus

Table 7.4. Summary of characteristic geometric parameters of lens system I. The symbol L denotes the distance of the centre of the lens element from the exit slit of the monochromator, B and H are the opening of the lens apertures in the horizontal and vertical plane, respectively, and D is the thickness of the lens element. Dimensions are in mm. The distance between the exit slit and the target (Fig. 7.6) is 60 mm

Element 1				Element 2				Element 3			
$L1$	$B1$	$H1$	$D1$	$L2$	$B2$	$H2$	$D2$	$L3$	$B3$	$H3$	$D3$
16	16∅	16∅	4	34	8	24	4	43	8	20	8

the lens system shown in Fig. 7.6 can be expected to have a large degree of flexibility in the vertical focusing. A summary of the geometric lens parameters is given in Table 7.4.

The more detailed results refer to the following arrangement: the distance of the object in the vertical plane from the exit slit of the monochromator is L_m = 150 mm and the distance of the target from the exit slit is 60 mm. As an illustration of the focusing we show in the vertical plane (Fig. 7.7a) a set of trajectories emerging with different angles β from the centre of the entrance slit of the (first) monochromator, which is displaced from the left hand boundary of the figure by L_m = 150 mm. In the horizontal plane Fig. 7.7b displays a bundle of trajectories emerging with different angles α from the centre of the exit slit. In both planes, a first-order image is achieved at the sample. The bottom part of Fig. 7.7 shows the shape of this image. The image is generated by plotting the points where the trajectories impinge on the sample. We have assumed slits of total height h = 4 mm, of total width s = 0.3 mm and an angular spread in α of

$$-\alpha_c < \alpha < \alpha_c \quad \text{with}$$

$$\alpha_c = \int_0^{\alpha_c} T_{\text{ideal}}(\alpha)\, d\alpha \,, \tag{7.33}$$

where $T_{\text{ideal}}(\alpha)$ is the transmission of the ideal cylindrical deflector as a function of the angle α with respect to the centre path at the entrance slit. For this transmission function we have found (3.29)

$$T_{\text{ideal}}(\alpha) = 1 - \frac{4}{3}\frac{r_0}{s}\alpha^2 \tag{7.34}$$

with the centre radius r_0 taken as 35 mm. Then α_c is

$$\alpha_c = \frac{2}{3}\sqrt{\frac{3s}{4r_0}} \,, \tag{7.35}$$

which is 2/3 of the transmitted angle α. We note that the aperture α_c at the entrance slit of the deflector is equal to the aperture angle of the bundle leaving the deflector and entering the lens system. The aperture angle β_c is determined by

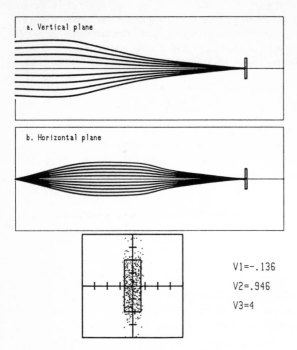

a. Vertical plane

b. Horizontal plane

V1=-.136
V2=.946
V3=4

Fig. 7.7a,b. Trajectories in lens I when the energy at the target is 4.3 eV and the energy at the exit slit of the monochromator is 0.3 eV. The voltages $V1$, $V2$, $V3$ refer to the potential at the exit slit and denote the voltage on elements 1, 2, and 3, respectively. The bottom part shows the image of the slit generated by 500 trajectories. The shape of the first-order image is also shown. One sees that part of the trajectories impinge on the sample outside the area of the first order image due to aberrations

the slit height, because the exit slit of the monochromator (or the last exit slit of a set of monochromators) act as a pupil in the vertical plane so that $\beta_c = h/2L_m$.

In addition to the positions of the electrons impinging on the sample, the area of the first-order image is shown as a rectangle (thick line in the bottom of Fig. 7.7). By comparing the shape of the rectangle with the dimensions of the slit ($h = 4$ mm, $s = 0.3$ mm), we realize that the aspect ratio has changed, which means that the magnification is substantially smaller in the vertical plane than in the horizontal plane. This is basically a consequence of the larger object-lens distance in the vertical plane (see also Fig. 7.3), as we have remarked earlier. An appreciable part of the trajectories fall outside the area of the first order image because of the third-order aberrations. The effect of aberrations is noticeable in particular in the vertical direction.

Focusing in the vertical and horizontal plane as schematically shown in Fig. 7.3 can be achieved over a wide range of impact energies at the target. In Fig. 7.8, the potential energies at the lens elements 1 and 2 necessary to obtain a focus at the target are plotted versus the energy at the target. We have plotted the potential energies rather than the potential applied in the simulation

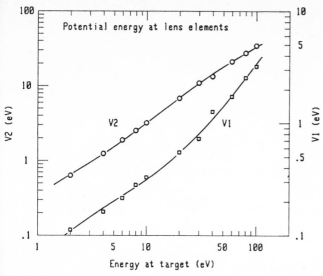

Fig. 7.8. Potential energy at the lens element 1 and 2 vs the target energy. When these potentials are applied, an image of the exit slit of the monochromator and the entrance slit of the first monochromator at $L_m = -150$ mm is formed in the horizontal and vertical planes, respectively

in order to avoid negative numbers in the plot. The initial kinetic energy was 0.3 eV in keeping with the monochromator layout (Sect. 5.1). The linear magnifications in the horizontal and vertical plane (M_x and M_y, respectively) are shown in Fig. 7.9. The magnification in the horizontal plane changes very little as a function of the impact energy, while the magnification in the vertical plane does. Instead of the magnifications one may also consider the acceptance angles at the sample in the two planes. In the horizontal plane the acceptance angle decreases from 1.54° to 0.421° when the impact energy is raised from 4 eV to 100 eV, while in the vertical plane the acceptance angle stays constant at 1.7°. These latter values refer to a monochromator/analyser slit height of 4 mm and the angle α_c defined above. The numbers are therefore not genuine properties of the lens. Nevertheless, for the purposes of illustration it may be useful to plot the $Q_\parallel$-resolution which follows from these acceptance angles. The result is shown in Fig. 7.10. Comparison with Table 7.2 shows that the errors in the determination of a frequency of an energy loss having a large dispersion are still acceptably small. We finally plot the significant third-order aberration terms in Figs. 7.11 and 7.12. These terms are the angular aberration and coma in the horizontal plane and angular aberration, coma and astigmatism in the vertical. Their calculation follows from the definitions of the quantities in (7.28) and (7.29) by choosing appropriate rays or combinations of rays and calculating the image positions. In that procedure, one assumes higher-order aberration to be small. The assumption may be checked by calculating the aberration terms for different α, β, x, and y, respectively and comparing the results. If higher-order aberrations are negligible the result should be constant. In practice, one finds that they are

135

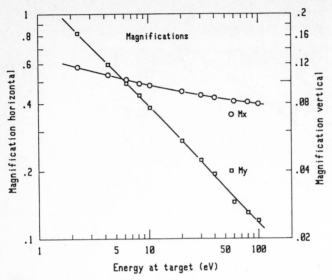

Fig. 7.9. Magnifications in the horizontal plane (M_x) and in the vertical plane (M_y) vs energy at the target for lens I

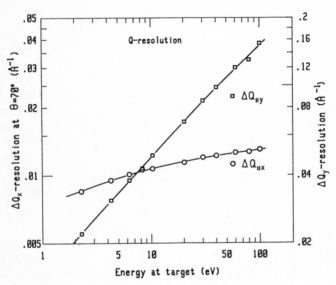

Fig. 7.10. $Q_{\parallel}$-resolution in the horizontal and vertical planes for lens system I. The $Q_{\parallel x}$-resolution in the horizontal plane depends on the angle of incidence (7.3) (here $\theta = 70°$)

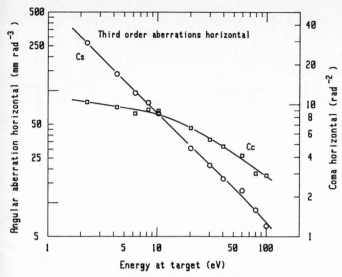

Fig. 7.11. Angular aberration and coma in the horizontal plane for lens I

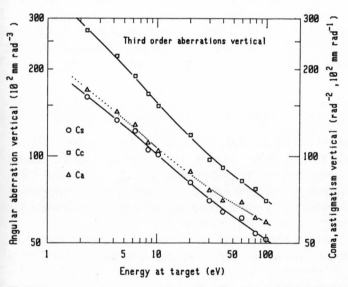

Fig. 7.12. Angular aberration, coma, and astigmatism in the vertical plane for lens I

Table 7.5. Third-order aberration terms in relation to the first-order image size as defined in (7.32). The quantities $\hat{C}_{sx}$, $\hat{C}_{cx}$, $\hat{C}_{ax}$, $\hat{C}_{dx}$ and $\hat{C}_{sy}$, $\hat{C}_{cy}$, $\hat{C}_{ay}$, $\hat{C}_{dy}$ are the reduced meridional aberrations with respect to the angle term, coma, the astigmatism, and distortion in the horizontal (xz) and vertical (yz) planes, respectively. The energy at the exit slit of the monochromator is 0.3 eV. The quantities refer to the maximum angle β_m defined by the pupil size of 4 mm (slit height) and a distance of the object from the exit slit of 150 mm. The maximum angle α_m is taken as $\alpha_m = \sqrt{3s/4r_0} = 4.59°$. The quantities given in this table are subject to several uncertainties and should therefore be considered as estimates only

Energy at target	$\hat{C}_{sx}$	$\hat{C}_{cx}$	$\hat{C}_{ax}$	$\hat{C}_{dx}$	$\hat{C}_{sy}$	$\hat{C}_{cy}$	$\hat{C}_{ay}$	$\hat{C}_{dy}$
4 eV	85%	8%	1%	11%	13%	34%	31%	8%
100 eV	5%	3%	1%	2%	27%	78%	81%	24%

not. One example may serve as an illustration: If for example, one calculates the coma error for $E_i = 100\,$eV in the vertical plane for a set of rays starting at $y = 2\,$mm at angles $\beta = 0.25°$, $0.5°$, and $0.75°$ one obtains for $C_{cy} = 70$, 100, and 145 rad^{-2}, respectively. This indicates that higher-order aberrations are not small. In Fig. 7.12 we have plotted the values obtained for the smallest angle, i.e. 70 rad^{-2}. It is obvious that the numbers generated in this manner are to be taken with a grain of salt. They are useful for the comparison of different lens systems, however. It is also illuminating to compare the various aberration terms in relation to the image size (7.31). In Table 7.5 the result is shown for two impact energies. The numbers are the average "third-order" aberrations calculated for x, y, α and β respectively. In keeping with the example, we therefore define

$$\hat{C}_{cy} = \frac{1}{3}(70 + 110 + 145)\frac{\beta_m^2 y_m}{M_y y_m} = 79\% \,, \tag{7.36}$$

where β_m denotes the maximum angle β determined by the maximum value of $y(= y_m = h/2)$ and the object distance. The other quantities are defined accordingly. Despite the fact that the precise values of these reduced distortions are affected by the way in which the calculation is performed, the comparison in Table 7.5 shows quite nicely which aberrations are important and which are not. The result changes with the impact energy. For a high impact energy, the aberrations in the vertical plane prevail, in particular coma and astigmatism, while for low impact energies the angular aberration in the horizontal plane becomes the largest aberration. Nevertheless, the transmission losses are due more to vertical aberrations than to horizontal (Fig. 7.7), presumably because of the sagittal aberrations. The sagittal aberrations are not evaluated here, since their calculation involves trajectories off the two symmetry planes. We expect the accuracy of the interpolation scheme and thus the accuracy of the trajectories to be worse there and quantitative numbers for the sagittal aberrations may thus be less meaningful.

A quantitative measure of the cumulative effect of all lens aberrations pertinent to the specific application we have in mind here is the fraction of the trajectories that arrive within the boundaries of the first order image. We call this fraction the "transmission" of a lens in the following. Like the reduced aberration coefficients discussed above, the transmission depends on the dimensions of the object (the slit) and the aperture angles of the beam. A qualitative comparison with an experimentally determined transmission is meaningful when a reasonable choice of the angular apertures is made. The quantitative numerical result does however depend rather critically on the shape of the angular distribution of the trajectories leaving the monochromator and also the distribution across the slit aperture. If the transmissions for two angular distributions of the same area but with different shapes (e.g. a rectangular distribution and a gaussian) are compared, the transmission of the broader distribution is lower since the part of the angular distribution in the wings of the distribution contributes with a smaller weight, if at all, to the transmission. Despite this drawback, the transmission as defined above is useful for the comparison of different lenses, when a suitable choice of the angular apertures and the slit dimensions is made. In the following, we assume the dimensions of the slit to be imaged to be $s = 0.3$ mm and $h = 4$ mm and the angular distributions to be rectangular in shape with cut-off angles $\beta_c = h/(2L_m)$; $L_m = 150$ mm and $\alpha_c = 2/3\sqrt{3s/4r_0}$ with $r_0 = 35$ mm as defined by (7.34). These values are the same as those for plotting the lower section of Fig. 7.7. Thus the transmission of lens I is the fraction of dots inside the boundary of the first order image drawn there. The value of the transmission for lens I at 4 eV is about 80%. This rather high transmission characterises a good lens. The transmission is also nearly constant for a wide range of impact energies so that one still has about 80% transmission at 100 eV target energy.

The transmission is also a useful quantity for an estimate of the chromatic error. This chromatic error is important when the analyser is used with an array of detectors rather than with an exit slit and a single detector. In that case, a multiplex gain is achieved only if the lens has a sufficiently small chromatic error: a wide energy distribution then enters the analyser and is spread along the array of detectors by virtue of the energy dispersion of the analyser. In Fig. 7.13 the transmission is plotted for lens I for two impact energies, 4 eV and 100 eV. The first order focus was adjusted to be at the sample position when the kinetic energy of electrons leaving the exit slit of the monochromator was 0.3 eV. One sees that a high transmission extends across a wide range of kinetic energies for 100 eV target energy. The transmission is a narrower function of the kinetic energy for a 4 eV target energy. The width is still large compared with the energy resolution of the analyser, so that parallel detection is feasible.

Because of the slow variation of the transmission with the kinetic energy, energy losses could be explored over a certain energy range even without readjusting the lens potentials, when one has a parallel detector at the end. For a single detector the energy loss range is scanned by raising the potential of the entire analyser to the value that corresponds to the energy lost in the scattering process from the sample. The resolution of the spectrometer is thereby kept constant. For optimum performance, the potentials applied to the two lens elements of the lens system between the sample and the analyser have to be varied also, in

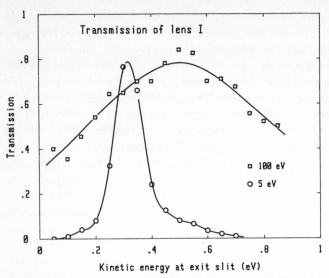

Fig. 7.13. Transmission of lens I vs the kinetic energy at the exit slit of the monochromator for impact energies of 4 eV and 100 eV at the target. The first order focus was adjusted to be at the target for a pass energy of the monochromator of 0.3 eV

order to keep the image of the electrons at the analyser entrance slit. The amount ΔV_1 by which the potential V_1 of element 1 of the lens is to be varied is a linear function of the amount by which the potential of the analyser is varied, which is equivalent to the energy loss ΔE_{loss}:

$$\Delta V_1 = M_f \, \Delta E_{\text{loss}} \, .$$

In German the proportionality constant is known as *Mitlaufverhältnis*. This term has been adopted by several laboratories in the shortened form as *Mitlauf*. For lens I discussed here, as for most other lenses, a Mitlauf of one is the best value for both lens elements as long as ΔE_{loss} is small compared with the energy at the sample.

We now discuss more briefly two further lens systems. The first employs a lens in which the beam can have a double focus in the horizontal plane and a single in the vertical. According to the calculations, this lens is inferior to the one described previously in nearly every way. The reason for mentioning this lens is that it was the first lens designed according to the focusing principles described at the beginning of this section (Fig. 7.3 and [7.8]) and has been in practical operation in this laboratory for several years. Despite being inferior in its electron optical properties, the lens has proven to perform quite acceptably in practice. Figure 7.14 shows the cross-section of this lens, referred to as lens II in the following.

The main difference between lenses I and II is that, instead of a round aperture for lens element 1, lens II exhibits a large extension of the aperture in the vertical

140

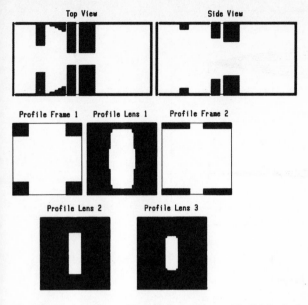

Top View   Side View

Profile Frame 1   Profile Lens 1   Profile Frame 2

Profile Lens 2   Profile Lens 3

Fig. 7.14. Cross-sections of lens II. See caption to Fig. 7.6 for further explanation

direction (see also Table 7.6). Consequently, one has weaker focusing in the vertical direction than with lens I for comparable potentials. Stronger focusing is achieved when the ratio of the potentials on element 2 and 1 is raised. In Fig. 7.15, the trajectories in the vertical and horizontal planes and the image points at the target are shown for 4.3 eV target energy. The figure should be compared with Fig. 7.7, which represents the trajectories with the same initial conditions and target energy for lens I. The potential on element 2 must now be made much larger for lens II: In fact it exceeds the target energy. In the vertical plane the trajectories arrive at the target with a larger angle for lens II, corresponding to a smaller magnification in the vertical plane ($M_y = 0.055$ instead of $M_y = 0.12$). In the horizontal plane the reverse is true and the magnifications in the horizontal plane are $M_x = 1.2$ and 0.53 for lenses II and I, respectively. As a consequence of these opposite variations in the magnifications, the shape of the first order

Table 7.6. Geometric parameters of lens system II. The symbol L denotes the distance of the centre of the lens element from the exit slit of the monochromator, B and H are the opening of the lens apertures in the horizontal and vertical plane, respectively, and D is the thickness of the lens element. Dimensions are in mm. The distance between the exit slit and the target (Fig. 7.6) is 60 mm

Element 1				Element 2				Element 3			
$L1$	$B1$	$H1$	$D1$	$L2$	$B2$	$H2$	$D2$	$L3$	$B3$	$H3$	$D3$
16	16	34	4	34	8	24	4	43	8	20	8

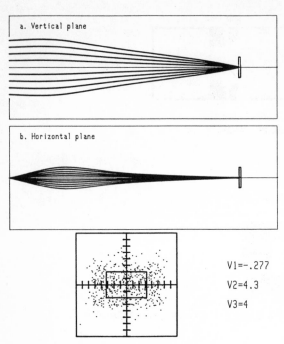

Fig. 7.15a,b. Trajectories for lens II in the single focus mode for a target energy of 4.3 eV. See caption to Fig. 7.6 for further explanation

image is quite different for lenses I and II. Because of the small magnification in the vertical plane for lens II, the image of the slit has a reversed aspect ratio with the height smaller than the width. Mostly as a consequence of the very small magnification in the vertical plane, the transmission of lens II is smaller ($\sim 40\%$). For high impact energies, lens II would need inconveniently high potentials on element 2. The double focus mode of operation shown in Fig. 7.16 is then preferable. In this mode both the lateral and the vertical magnifications of lens II become comparable with those of lens I. (The first order images in Figs. 7.6, 15, and 16 are not drawn in the same absolute scale). The transmission drops further to about 35%. In the double focus mode, lens II can be used for rather high impact energies, the potentials on element 2 remaining conveniently small: for a target energy of 500 eV only about 30 eV are necessary. The disadvantage is the lower transmission. Lens II was also found to be more sensitive to spurious potentials on the electrodes making up the lens element 1 and the frames connected to lens element 1. This is because the kinetic energy of the electrons in this lens element becomes rather small in the double focus mode. For the same reason, lens II has a large chromatic error and performs in fact as a band-pass filter in the double focus mode.

The last lens to be discussed in this section has, except for slightly different shielding frames and dimensions of the apertures (Table 7.7), the same lens components as lens I. In contrast to lens I, the entire package of lenses can be

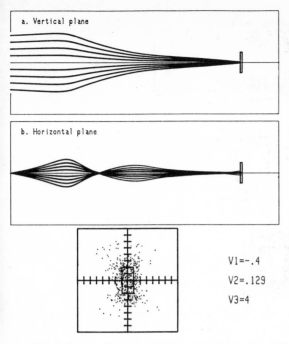

a. Vertical plane

b. Horizontal plane

V1=-.4
V2=.129
V3=4

Fig. 7.16a,b. Trajectories for lens II in the double focus mode in which the lens can operate with a moderately low potential on element 2 even at high target energies

moved to a position close to the target, which is now placed at a distance of 105 mm from the exit slit instead of 60 mm. The lens is shown in Figs. 7.17 and 7.18 for the two extreme positions. The corresponding trajectories again for 4.3 eV target energy and 0.3 eV initial energy are displayed in Figs. 7.19 and 7.20 respectively. In both cases the transmission is reasonably high ($\sim$ 80% and $\sim$ 70%, respectively). The difference between the two positions of the lens package is of course the different magnification (Figs. 7.21 and 7.22) and consequently the different aperture angles at the target. Depending on the impact energy, the magnification, and hence the aperture angles, can be varied by an order of magnitude. In the high angular resolution mode with the lens package as far from the target as possible, the angular resolutions become such that angle resolved studies of dipole losses should become feasible.

Table 7.7. Geometric parameters of lens system III. The lens package is movable by the amount $0 < L_x < 58$ mm. Definition of the symbols as in Table 7.4

Element 1				Element 2				Element 3			
$L1$	$B1$	$H1$	$D1$	$L2$	$B2$	$H2$	$D2$	$L3$	$B3$	$H3$	$D3$
$4 + L_x$	160	160	4	$22 + L_x$	10	24	2	$32 + L_x$	10	24	8

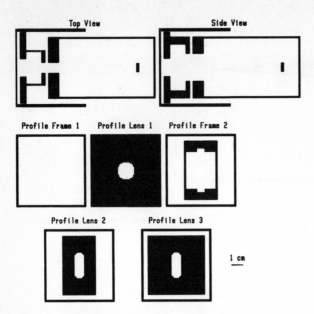

Fig. 7.17. Cross-sections of lens III, when the package of lens elements is close to the exit slit of the monochromator and the distance to the sample on the right-hand side of this figure is largest

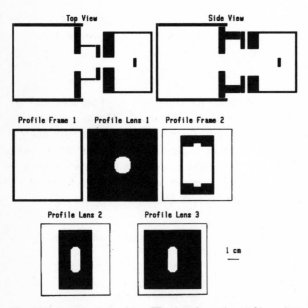

Fig. 7.18. Cross-sections of lens III, when the package of lens elements is moved close to the target. In practice, the shortening of the target chamber is achieved by folding pieces of sheet metal around two movable and one fixed pivot. The different shape of the target chamber in that case is of no concern because the target chamber is a field-free space

144

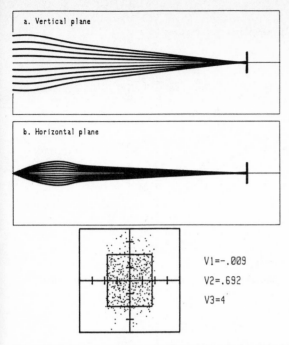

a. Vertical plane

b. Horizontal plane

V1=-.009
V2=.692
V3=4

Fig. 7.19a,b. Trajectories in lens III when the lens package is near the monochromator. The trajectories arrive at the target with a small angular aperture (high angular resolution mode). The potential on frame 1, which shields the area between the monochromator and the lens elements 1, is kept at the mean of the potential of the exit slit and that of element 1

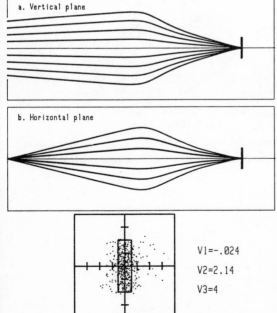

a. Vertical plane

b. Horizontal plane

V1=-.024
V2=2.14
V3=4

Fig. 7.20a,b. Trajectories in lens III in the close-up position yielding the highest intensity at the expense of low resolution in angle and $Q_{\parallel}$-space

145

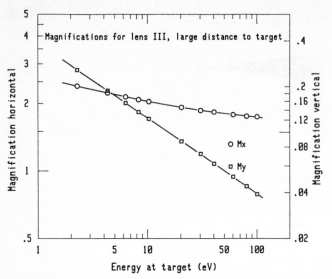

Fig. 7.21. Horizontal and vertical magnification of lens III in the high angular resolution mode

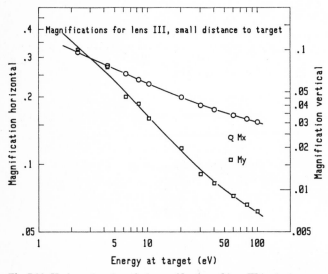

Fig. 7.22. Horizontal and vertical magnification of lens III in the low angular resolution mode

7.4 The Pupils of the Spectrometer

We have now considered all the electron optical elements of an electron spectrometer save one, the element that acts as a pupil in the horizontal plane. In the vertical plane, we have already seen that with the focusing shown in Fig. 7.3 the exit/entrance slit of the monochromator/analyser serves as the pupil that limits the maximum angle β in the vertical plane. In the horizontal plane, the only physical defined aperture that could eventually limit the maximum angle α would be the deflecting plates of the monochromator and analyser. If, however, these deflecting plates were indeed the effective pupil, angles α_m as large as 20°–30° would be permitted in the deflectors. Due to the second order aberration term (3.4, 4.16), energies quite distant from the nominal pass energy would still be transmitted by the deflectors; in other words the spectrometer would have a uselessly low resolution if there were no pupil restricting the angle α to a much smaller value. We have already discussed this matter in detail in Sect. 3.3. For the monochromator, a properly designed cathode emission system as described in Chap. 6 can provide a feed beam to the monochromator with a sufficiently small aperture angle α_m. Since small aperture angles α_m lead to a high transmission of the monochromator and consequently to a high monochromatic current, tuning up a spectrometer to provide high currents means (among other things) that one is also tuning the emission system to provide a feed beam with a small aperture angle. This is in particular true when the performance of the spectrometer is observed with the current at the detector. This ensures that one is tuning the monochromator and the cathode so that they produce the highest possible *monochromatic* current, not just some current with unspecified monochromaticity. In practical experiments, we found that tuning as described above was always sufficient to give a resolution of the monochromator close to the theoretical value.

When the monochromatic current is directly observed with analyser and detector with no scattering from a sample, the aperture angle of the current at the target as defined by the emission system also defines the aperture angle for the analyser system and a high resolution of the entire system can be achieved even with no explicit pupil in the analyser section. This is no longer true when electrons scattered from a sample are analysed. In general, the losses have a wider angular distribution and a beam of a wide angular distribution would hence enter the analyser. This is the explanation for a very typical observation, namely the resolution observed in the energy losses and with elastically diffuse scattered electrons is lower than with the direct beam. The degrading of the resolution is however rarely disastrous, typically 10%–20%, which is much less than one might expect considering the spread of angles α the analyser could let pass between the inner and outer deflection plate. The reason for this rather fortunate result can be understood by considering the aberration terms containing the angle α of the lens between the sample and the analyser. As a consequence of these aberration terms (angular aberration, coma and astigmatism, but mostly the angular aberration), electrons leaving the sample with larger angles α from the centre of the sample area illuminated by the primary beam cannot enter the entrance slit of the analyser. By the same token, electrons that leave the sample at the same large angle α but from a different point could pass through the slit. That part of the sample area is however not illuminated, and hence no scattered electrons emerge from there. Thus we see that angular aberrations sometimes offer certain advantage. On the other hand, it is dangerous to rely on such ill-deffined effects when resolution is concerned. We recall that losing a factor of two in the resolution is equivalent to an order of magnification drop in the signal at the detector. It is therefore advisable to have a well-defined pupil and restrict the maximum angle of the analyser. This is more important the *better* the lenses are! If one has a single detector, a continuous dynode multiplier for example, one may place a second slit after the exit slit of the analyser to restrict the maximum angle. An additional benefit is that electrons scattered from the deflecting plates are prevented from entering the detector. Such electrons can appear as ghost peaks in the spectrum [7.12]. A specific example of such a slit which also serves as a lens to focus the electrons into the multiplier is shown in Fig. 7.23, together with electron trajectories emerging from the center of the exit slit in the horizontal plane and from a point placed 150 mm to the left in the vertical plane. The width of the slit lens was chosen to be 2 mm at a distance of 4 mm from the exit slit, blocking geometrically angles larger than 14°. With the applied negative bias on the slit a focus is obtained in the horizontal plane at the opening of the multiplier. Due to this bias the cutoff angle α is reduced to about 10°. By lowering the potential on the slit further, the cutoff angle can be continuously reduced to zero, while the beam in the horizontal plane remains confined to a small area in the

Channeltron Lens

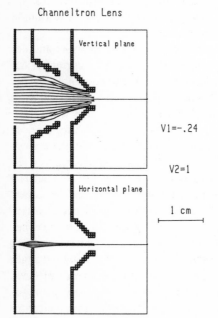

V1=−.24

V2=1

1 cm

Fig. 7.23. Lens aperture between the exit slit of the analyser and the funnel of a continuous dynode multiplier placed behind a circular hole in a second plate. The potential on this plate and the funnel is 1.3 eV, the energy of the electrons at the exit slit is 0.3 eV. $V1$ and $V2$ denote the potential differences between the exit slit and the slit lens and the multiplier, respectively. By adjusting the voltage $V1$ at the slit lens, the maximum angle α_m transmitted in the horizontal plane can be varied continuously down to zero

center of the multiplier. The focus shifts towards the slit, however. In the vertical plane, a pair of guiding plates as shown in Fig. 7.23, which extend across the entire slit plate, provide some focusing so that all electrons emerging from a slit of 12 mm height can be guided into the active area of the multiplier funnel. With such a slit lens, the analyser is equipped with a defined, adjustable pupil which adds the final touch to the performance of the spectrometer, because it secures the resolution independently of the status of lenses, the scattering processes at the sample and the nature of the sample surface.

8. Comparison of Experiment and Simulation

This chapter provides further details of a particular spectrometer design. Experimentally determined parameters such as current, resolution, transmission and acceptance angles at the target are compared to results from numerical simulations, and nearly perfect agreement is found. The performance of the spectrometer is demonstrated with the vibration spectrum of a monolayer of CO featuring a resolution of better than 1 meV.

8.1 Layout of the Spectrometer

In the course of the development of the computer codes for the various electron optical elements, as described in the previous chapters, the comparison with experimental results was rather crucial. While the accuracy and convergence of the results of the simulation can be checked by computational methods alone, the specification of the parameters for the beam entering each optical element does in general require comparison with experiment. Some of these parameters, such as the energy spread, can be determined directly in an electron spectrometer. Other parameters, however, such as the angular spread of the beam when it enters the first monochromator, are not amenable to experimental determination with a reasonable effort. Only through a detailed comparison of theoretical and experimental results can one obtain a feeling for the magnitude of those parameters and build up experience. This experience has of course influenced the specifications of the optical elements described in the previous chapters. Furthermore experimental tests are indispensable for the purpose of optimising a multi-parameter system in which the number of design parameters and voltages is of the order of a hundred, since the full exploration of the parameter space would be beyond the capacity of even large computers. Rather than describing the comparison of experimental and theoretical results of each electron optical element independently as in the earlier chapters, we devote this last chapter to the comparison of theory and experiment for all parts of a spectrometer. We are thus in a position to describe the layout and performance of a rather advanced electron energy loss spectrometer, which was designed with the experience and insight supplied by the computer simulation. The design of this spectrometer is described in the following.

For the electron emission system, we have selected the slot lens system (Fig. 6.4). The choice was based on the fact that the radiance of the slot lens system is reasonably independent of the size of the cathode tip (Fig. 6.17) and that the slot lens system provides an independent control of vertical and horizontal focusing. The slot lens emission system requires a more accurate adjustment of the potentials than the circular lens system. Quite acceptable spectrometer performance was likewise obtained with the circular lens system, which is less critical in the potentials. The circular lens system is particularly recommended when digitally controlled, independent power supplies are either not available or planned.

The first monochromator of the spectrometer operates as a retarding monochromator with a projected retardation ratio of about 1 : 5. The retardation is intended to be "exit aperture retardation", that is, it takes place near the exit aperture of the pre-monochromator. This exit aperture is also the entrance aperture of the second monochromator. The potential of the entrance aperture of the pre-monochromator is independently controlled, so that retardation right after the entrance aperture can be employed, when desirable (Sect. 5.2). The radii of the inner deflection plate, of the centre path, and of the outer deflection plate were chosen to be 25 mm, 35 mm and 45 mm, respectively. The total height of the deflector is 44 mm with a 2 mm gap between the deflection plates and the top and bottom cover plates. Such a deflector has, in the absence of space charge, a first-order focus at a deflection angle of approximately 109° when the potential applied to the top and bottom cover plates eU_D is such that the trajectories are straight lines in the vertical plane, which is the case when $eU_D \sim 0.22\, E_0$, E_0 being the nominal pass energy defined by (3.1) and (3.2). The value is dependent on the retardation ratio but is practically independent of the actual deflection angle, extended or not. As a consequence of space charge, the beam tends to spread in the vertical plane. This spreading should be compensated by applying a negative bias to the top and bottom cover plates. As shown in Fig. 3.9 such a negative bias increases the deflection angle, leaving less room for further extension of the first order focus due to space charge. In the course of our simulation we found that the pre-monochromator studied in Sect. 5.2 with a total deflection angle of 127° failed to provide enough current to feed a second nonretarding monochromator, when the latter was built with a deflection angle of 114°, as studied in Sect. 5.1. For our new design we have therefore extended the total deflection angle of the pre-monochromator to 140°. With this deflector the trajectories in the vertical plane remained straight lines when the input current was adjusted to have the first order focus at 140° and when the cover plates were negatively biased to $\approx -0.55\, E_0$. Here E_0 is again the nominal pass energy calculated from (3.1) and (3.2). In the absence of space charge, this particular bias would focus a beam of parallel trajectories in the vertical plane approximately at the exit slit. The focusing angle in the horizontal plane in the absence of space charge is about 119°, i.e. 8° larger than without the negative bias.

The magnification of this deflector under space charge conditions was found to be $C_y = -1.6$, approximately in accordance with (3.18). Despite this magnification we have abstained from reducing the entrance slit with respect to the exit slit. The reason is that the performance of the second monochromator is only weakly dependent on the resolution of the first monochromator. On the other hand the entrance slit is loaded with a large flux of electrons and negative charging of the aperture plate near the slit can probably not be avoided. This negative charging reduces the effective width of the slit and could also hinder the formation of a well defined entrance beam. In the numerical simulation, we found the half-width of the output beam to increase by 25% when the width of entrance slit (0.3 mm) and exit slit were made equal compared with a pre-monochromator where the entrance slit was reduced according to the magnification. The simulation did not of course include charging effects on the slit. In accordance with the expected shape of the feed beam (Figs. 6.7, 8), we have made the entrance slit 3 mm high, while the exit slit allowed for 6 mm height of the emerging beam. A final comment with regard to the first monochromator layout is the offset of the radial position of the exit slit with respect to the entrance slit of about 1.5 mm, which ensures that the bundle of emerging trajectories is centred around $\alpha_2 = 0$ when the entering trajectories are also centred around $\alpha_1 = 0$ (see Sect. 3.2).

The second monochromator had the same dimensions as the first, apart from a reduction in the deflection angle to 114° in accordance with the considerations in Sect. 5.1. Since the second monochromator is supposed to operate as a nonretarding deflector, there is no radial offset of the exit slit. Both monochromators and the analyser have a saw-tooth profile milled into the inner and outer deflecting plates, so that electrons striking the surface near grazing incidence are scattered backwards rather than being reflected in the forward direction. This rather effectively reduces the background of electrons appearing in a spectrum at unexpected energies.

As the lens system between the monochromator and sample and between the sample and analyser, we use the rather flexible lens depicted in Figs. 7.17 and 18. The stack of lens elements 1–3 was mounted on a linear ball-race. The linear motion was controlled by a system of rope and pulleys, such that the distances between the sample and the two lenses were always kept equal. Thus the basic symmetry of the lenses as discussed in Sect. 3.3 was ensured. The entire package comprising the emission system, the two monochromators, and the lens guiding the beam to the sample is mounted on a rotatable table with the pivot on the optic axis at the sample position.

A short note on material selection and on some details of the manufacturing process may be helpful. Lens elements were made from OHFC copper and coated with a suspension of collodial graphite in isopropanol (Dag 154, Acheson). Copper is preferred over aluminium or molybdenum because its oxide is conductive, not insulating, unlike the others. We experienced that a monochromator made from molybdenum failed to operate after several years of service because of oxidation and the resulting surface charging effects. The cylindrical deflectors have tungsten wire heaters for a separate bakeout, typically performed after the bakeout of the main vacuum chamber, while still at elevated temperature.

Mechanical parts other than the lens elements were made from truly antimagnetic stainless steel (DIN 1.3952). Screws and ball bearings from merely nominally antimagnetic steel proved to be inadequate in their magnetic properties and were replaced by copper-beryllium screws and sapphire balls, respectively. Commercially available manipulators sometimes require the replacement of magnetic parts. It should be considered that the highly inhomogeneous magnetic field stemming from small parts within the system is much more devastating to the performance of the spectrometer than a possible residual homogeneous field. The same is true for ac fields. In order to reduce all components of the magnetic field, including the inhomogeneous and ac parts, the entire chamber is shielded magnetically by a cylinder of high permeability material ("conetic"). The wall thickness should be about 5×10^{-3} of the diameter of the cylinder, so that the residual magnetic field is below $10 \, \text{mG}$ and the inhomogeneity smaller than $1 \, \text{mG/cm}$. If the spectrometer is placed inside the cylinder such that the distance from the opening at one end is larger than the diameter of the cylindrical shield, then the end of the cylinder can remain open, which is in the interest of a large pumping speed. We finally remark that the electrons must not "see" any insulating material. Where this cannot be avoided the insulating material must be placed inside a cavity of conducting material that is electrically connected to appropriate potentials.

In previous designs we have frequently used an analyser consisting of two deflectors of 35 mm radius. The purpose of the second stage was to reduce the background of electrons scattered from the deflection plates. In particular, when one probes an energy loss range for which the beam of elastically scattered electrons strikes the outer deflection plate, a considerable fraction of the electrons is reflected from the plate and can pass through the exit slit, despite their "wrong" energy. These "ghost peaks" can be suppressed very effectively by using a two-stage analyser. In addition to having the wrong energy, scattered electrons also leave the exit slit at a large angle. This is particularly true when the gap between the inner and outer deflection plate is large. An aperture placed behind the exit slit can therefore also prevent scattered electrons from entering the detector. A combination of "saw-tooth" profiling of the deflection plates, a wide gap between the deflection plates, and an exit aperture lens (Fig. 7.23) is therefore efficient enough to reduce the background electron level to an unmeasurably low value, below the dark count rate of the channel electron multiplier ($< 1 \, \text{s}^{-1}$). In Sect. 3.3 we have shown that the analyser should have a better resolution than the monochromator, and we have estimated that an optimum match between analyser and monochromator is achieved when $\Delta E_{1/2A} \sim \sqrt{2/5} \Delta E_{1/2M}$ (3.53), where $\Delta E_{1/2A}$ and $\Delta E_{1/2M}$ are the full width at half maximum for the analyser and monochromator, respectively. Since the monochromator has a relative resolution of $\Delta E_{1/2}/E_0 \sim 9.7 \times 10^{-3}$ (Fig. 5.5), a relative resolution of $\Delta E_{1/2}/E_0 \sim 6.1 \times 10^{-3}$ would be indicated for the analyser. We recall too that the exit slit of the monochromator and its first-order image at the entrance aperture of the analyser are of the same size when the pass energies of monochromator and analyser are the same. The required larger resolution of the analyser therefore calls for a larger centre radius. Since aberrations may enlarge the image of the exit slit of the monochromator at the analyser entrance, an enlargement of the entrance (and exit) slit of the analyser seems advisable. We have therefore made the entrance and exit slits of the analyser of 0.4 mm width and 10 mm height and the centre radius $r_0 = 70$ mm, the latter in keeping with the required relative resolution. All the other dimensions of the analyser are also scaled up by a factor of two compared to our reference deflector with 35 mm radius. The optimum, zero-current deflection angle is thus again 107°.

The potentials applied to the spectrometer are provided by digitally controlled, independent power-supplies. Each power-supply consists of a 16-bit digital-to-analog converter (DAC), which in turn drives a low-noise, low-drift and low-offset operational amplifier (OP). Since many potentials are required in pairs, a DAC typically drives two OPs at a time. The temperature drift of the potentials is about $30 \, \mu V/°C$ and the peak-to-peak noise level is below $200 \, \mu V$. This low noise level is ensured by wide-band AC-filters on each power supply feeding the DAC-OP combinations and additional HF-filters in the exit line of each potential. The operational amplifiers also feature a second analog input, which is used to build up a master-slave circuitry. In the scan mode of the spectrometer, where many potentials need to be changed simultaneously, only one potential is digitally ramped by the computer, while the others follow in the analog slave mode. A dedicated keyboard serves to address each individual potential by a push button and a set of three digital ramps with different speeds is used to adjust the individual potentials in the procedure of optimising the intensity. The current can be measured at all apertures of the two monochromators and at the entrance aperture of the analyser. This allows sequential optimisation of all parts. It should be mentioned, however, that the only relevant final test is the measurement of the current *and* the resolution at the detector, since only when optimising the current in a narrow energy window does one optimise the transmission of the monochromator and lens system. Such an optimisation also automatically takes care of the optimum choice of angular aperture of the beams feeding the first and second monochromator and the right choice of current level for each of the space charge compensated monochromators. The greatest advantage of the computer-controlled power supply of the spectrometer is that after each recorded spectrum the entire set of potentials can be stored automatically together with the data, so that the potential setting can be recalled at any later time. Even after intermediate bakeout of the vacuum chamber, we found that calling-in previous potential sets led to a reasonable beam intensity at the electron detector with only minor further adjustments needed for optimisation.

8.2 The Analyser

We now turn to the experimental analysis of the electron optical elements of the spectrometer described above and begin our examination with the analyser. The spectrometer potentials are initially set so as to optimise the monochromatic current at the electron multiplier (used as a Faraday cup), with a pass energy of 0.3 eV in the monochromator and 1.8 eV in the pre-monochromator. The energy at the target position was set to 4 eV and the lens packages were moved to positions close to analyser and monochromator (compare Fig. 7.17). The monochromatic current at the detector was found to be about 0.1 nA at 4 meV resolution (FWHM) of the spectrometer.

The first test on the analyser involves the determination of resolution as a function of the nominal pass energy, or more conveniently, of the potential difference ΔU_A between the deflection plates. According to the numerical calculation of the dispersion (3.17) and the full width at half maximum of the energy distribution $\Delta E_{1/2A}$, one should have (3.56)

$$\Delta E_{1/2A} = \frac{\Delta U_A}{2\ln(R_2/R_1)}\left(\frac{s}{0.966\,r_0} + 0.47\,\alpha_{1m}^2 + 0.6\,\beta_{1m}^2\right)\;. \tag{8.1}$$

Assuming gaussian profiles for the transmitted energy distribution, which is a good approximation, the FWHM of the entire spectrometer $\Delta E_{1/2\,tot}$ should be

$$\Delta E_{1/2\,tot}^2 = \Delta E_{1/2\,M}^2 + \Delta E_{1/2\,A}^2\;. \tag{8.2}$$

Plotting $\Delta E_{1/2\,tot}^2$ vs ΔU_A^2 should then give a straight line

$$\Delta E_{1/2\,tot}^2 = \Delta E_{1/2\,M}^2 + \text{const} \times \Delta U_A^2\;. \tag{8.3}$$

This is indeed the case, as shown in Fig. 8.1. The result confirms the assumption of gaussian profiles and the consequent geometric additivity of the full width at half maximum of the convolution of analyser and monochromator transmission curves, which is the transmitted energy spectrum of the spectrometer. From Fig. 8.1 we take $\Delta E_{1/2\,M} \sim 3.2\,\text{meV}$ as the resolution of the monochromator. This figure compares well with the results of the numerical simulation of the space charge compensated deflector in Chap. 5 (Figs. 5.5, 6, 10). The details of the monochromator performance are discussed in the next section.

The constant slope of the line in Fig. 8.1 could be calculated from (8.1) if the maximum input angles α_{1m} and β_{1m} were known. Since the monochromator was adjusted to optimum current in a small energy window (as provided by the analyser), we can expect the angular aperture of the beam emitted from the monochromator to be about 3° (Fig. 5.8). By virtue of the reciprocity of the lens system, one expects the angular aperture of the beam entering the analyser to be about the same. The maximum angle β_{1m} can be estimated from the maximum deviation in β from the centre path allowed by the slit height of 10 mm which

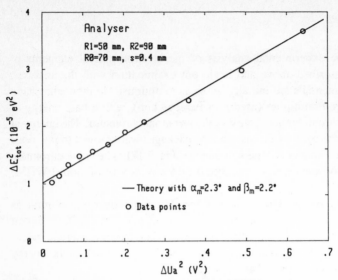

Fig. 8.1. Square of the total energy resolution (FWHM) vs the square of the potential difference between the deflection plates of the analyser. The data follow the straight line $\Delta E^2_{1/2\,\text{tot}} = (3.2\,\text{meV})^2 + 4.08 \times 10^{-5}\,\text{V}^{-2}\Delta U^2_{\text{A}}$ in accordance with (8.1–3)

results in $\beta_{1m} \approx 2.2°$. Figure 8.1 shows a fit to the data assuming $\beta_{1m} = 2.2°$ and $\alpha_{1m} = 2.3°$.

The transmission of the analyser is defined as the fraction of the current leaving the analyser relative to the current entering the analyser. The theoretical transmission has been calculated in Sect. 3.3. The expression contains two factors, one depending on the angular aperture of the feed beam (3.30) while the other takes the energy window into account (3.48). The total transmission is therefore

$$T\left(\alpha_{1m}, \Delta E_{1/2\,\text{A}}, \Delta E_{1/2\,\text{M}}\right) = \frac{\left[1 - (|C_{\alpha\alpha}|/3s)\alpha^2_{1m}\right]}{\left[1 + \left(\Delta E_{1/2\,\text{M}}/\Delta E_{1/2\,\text{A}}\right)^2\right]^{1/2}} \cdot \qquad (8.4)$$

In Fig. 8.2 the experimental results are compared with this theoretical expression when $\alpha_{1m} = 2.3°$ (the value obtained from matching the resolution). The angular aberration $C_{\alpha\alpha}$ is taken as $-1.48\,r_0$ (3.16), r_0 and s are 70 mm and 0.4 mm. The agreement between the theoretical transmission and the experimental data (full line in Fig. 8.2) is quite close. We can therefore conclude that the resolution and transmission of the analyser are according to expectation. It is useful to determine the optimum setting of the analyser resolution relative to the monochromator resolution. In Sect. 3.3 we have already addressed this issue theoretically and derived an expression for the monochromatic current at the detector as a function of the ratio $x = \Delta E_{1/2\,\text{A}}/\Delta E_{1/2\,\text{tot}}$ for two cases: where the monochromatic beam is measured directly or where one samples electrons emerging from a target with a diffuse angular distribution. We found

156

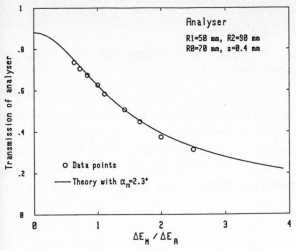

Fig. 8.2. Transmission of the analyser versus $\Delta E_{1/2\,M}/\Delta E_{1/2\,A}$ compared with the theoretical transmission with $\alpha_{1m} = 2.3°$ and $\beta_{1m} = 2.2°$

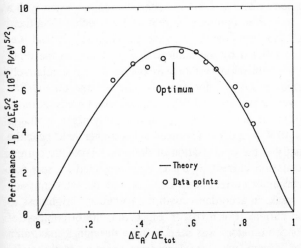

Fig. 8.3. Performance factor $I_D/\Delta E_{tot}^{5/2}$ of the spectrometer as a function of the ratio $x = \Delta E_{1/2\,A}/\Delta E_{1/2\,tot}$ together with the theoretical curve $\sim x(1 - x^2)^{5/4}$, which is discussed in the text (8.5)

$$\frac{j_D}{\Delta E_{1/2\,tot}^{5/2}} \sim x(1 - x^2)^{5/4} , \qquad \text{direct beam} , \tag{8.5}$$

$$\frac{j_D}{\Delta E_{1/2\,tot}^{7/2}} \sim x^2(1 - x^2)^{5/4} , \qquad \text{diffuse scatterer} . \tag{8.6}$$

The theoretical curve corresponds to the data as obtained for the direct beam quite closely (Fig. 8.3) with a soft optimum at

$$\Delta E_{1/2\,\mathrm{A}} = \sqrt{\frac{2}{7}}\,\Delta E_{1/2\,\mathrm{tot}} = \sqrt{\frac{2}{5}}\,\Delta E_{1/2\,\mathrm{M}} \ . \tag{8.7}$$

Taking the resolution of the analyser and the monochromator into account, our particular spectrometer performs best when

$$\Delta U_{\mathrm{A}} \sim 0.9\,\Delta U_{\mathrm{M}} \ . \tag{8.8}$$

In summary we find that the analyser does perform according to the theoretical analysis. Although the data in Fig. 8.3 agree well with (8.5), we must not forget that the monochromatic current should be proportional to $\Delta E_{1/2}^{5/2}$. While we shall find this to be approximately the case for an energy width of the monochromatic beam around 3–5 meV, we shall see shortly that the $\Delta E_{1/2}^{5/2}$ dependence is not a universal one.

8.3 Emission System and Pre-monochromator

The properties of the emission system and the monochromators are tested with a spectrometer adjusted to produce an optimum current at the detector for a fixed resolution on the one hand, and with potentials on the emission system and pre-monochromator chosen as predicted by the numerical simulation, on the other hand. In Chap. 6 the slot lens emission system was operated with a relatively small field near the cathode. In Fig. 6.8 for example, the voltage on the first lens element and repeller were 2 V and −0.605 V, respectively. Optimisation of the monochromatic current at the detector leads to substantially higher voltages on these two elements ($\sim$ 40 V, −2.6 V). The need for a higher field near the cathode stems from the fact that the optimisation of the monochromatic current also called for a higher emission current (6–8 μA) than projected in Sect. 6.4. The increase in the monochromatic current at the detector with the total emission current is however rather weak, in accordance with the calculated brightness as a function of the emission current (Fig. 6.16). In Fig. 6.16a, the brightness (of the beam entering the monochromator) was seen to pass through a maximum in the case of the slot lens system, when the lenses are operated in the realm of low voltages at the first lens element and repeller. Our experimental results suggest that the maximum can be shifted to higher emission currents and to higher brightness levels, when the voltages are readjusted to produce a higher field near the cathode tip. Since the result appears to be reasonable, we did not reinvestigate the issue theoretically, in particular since our calculations of the trajectories converge rapidly only for moderate space charge.

A fundamental property of the emission system is the energy distribution of electrons injected into the entrance slit of the pre-monochromator. As shown theoretically, this energy distribution is considerably smaller than the Maxwellian distribution emitted from the cathode due to the chromatic error of the cathode lens system. The calculated energy distributions in Fig. 6.13 referred to a

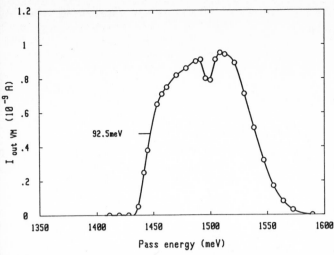

Fig. 8.4. Output current of the pre-monochromator vs the pass energy, measured as the current falling onto the exit aperture of the second monochromator. The dip in the centre of the curve occurs when the pass energy of pre-monochromator and second monochromator match and a fraction of the beam leaves the second monochromator through the exit slit. Apart from the dip, the curve represents the energy distribution of electrons entering the pre-monochromator

potential of 1.5 eV at the entrance slit and a slit width of 0.15 mm. We now compare this theoretical result (which included the effect of space charge) to an experimental measurement. For this purpose, we recorded the current falling onto the exit aperture of the monochromator as a function of the pass energy of the pre-monochromator, defined by the potential difference applied to the deflection plates of the pre-monochromator. The result is shown in Fig. 8.4. The current measured at the exit slit of the second monochromator is equal to the current into that monochromator, except when the energy of the electrons matches the pass energy of the second monochromator, in which case the electrons leave the second monochromator through the exit slit. This causes the dip in the energy distribution in Fig. 8.4. The depth of the dip is equal to the monochromatic current leaving the second monochromator, which is about 0.2 nA. The ratio of the areas under the dip and under the entire curve is a measure of the transmission of the monochromator to which we shall return in Sect. 8.4. Here we note that the width of the energy distribution in Fig. 8.4 is ∼ 92.5 meV. The energy distribution represents the convolution of the energy distribution entering the pre-monochromator with the energy transmission function of the pre-monochromator. The latter being rather small (as will be discussed shortly), the full width at half maximum of the distribution in Fig. 8.4 represents the width of the distribution of the beam entering the pre-monochromator to a very good approximation. Thus the value of $\Delta E_{\text{in VM}} = 92.5$ meV is to be compared directly with the theoretical value of ∼ 120 meV in Fig. 6.13. Given the approximation in the calculation of the lens properties, the match between experimental and theoretical result is

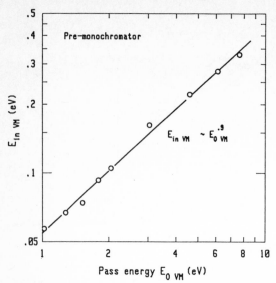

Fig. 8.5. Experimentally determined width (FWHM) of the energy distribution of electrons entering the entrance slit of the pre-monochromator vs the potential of the entrance slit

rather pleasing. Because the numerical simulation is rather time consuming, we did not investigate all the properties of the emission system as a function of the applied potentials. In particular, we did not study in detail the width of the energy distribution as a function of the slit potential under space charge conditions. There is, however, some evidence that the small energy width of the beam entering the pre-monochromator at low pass energies stems to a good part from the effect of space charge in the realm of deep space charge conditions. Our experimental investigation showed the width $\Delta E_{\text{in VM}}$ to be nearly proportional to the slit potential in the range of interest here (Fig. 8.5).

The dependence of the width of the energy distribution of the feed beam of the pre-monochromator on the pass energy is of fundamental importance for the monochromatic current produced by the pre-monochromator. From (5.15) we can now expect the monochromatic current to be proportional to $\Delta E_{1/2}^{3/2}$ rather than to $\Delta E_{1/2}^{5/2}$ where $\Delta E_{1/2}$ is the FWHM of the beam leaving the pre-monochromator. Experimentally, the width of the energy distribution of electrons leaving the pre-monochromator was determined by observing the current emerging from the second monochromator (at the entrance aperture of the analyser, beam defocused) as a function of the difference between the pass energies of monochromator and pre-monochromator. The resulting energy distribution is again the convolution of the energy transmission curves of pre-monochromator and monochromator. The energy width of the beam emerging from the pre-monochromator is then calculated assuming gaussian transmission functions for both monochromators. The result is plotted in Fig. 8.6. In order to obtain the data in Fig. 8.6, the pass energies of both monochromators were varied proportionally, so that the ratio of the resolutions of the dispersing elements remained approximately constant. Figure 8.6 shows that the resolution of the pre-monochromator $\Delta E_{1/2\,\text{VM}}$ is not

160

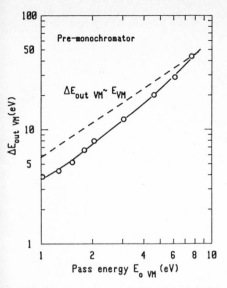

Fig. 8.6. Energy resolution (FWHM) $\Delta E_{1/2\,\mathrm{out\,VM}}$ of the pre-monochromator vs the pass energy $E_{0\,\mathrm{VM}}$ for a constant exit aperture retardation factor of 5. The dashed line represents the expected proportionality of $\Delta E_{1/2\,\mathrm{out\,VM}}$ and $E_{0\,\mathrm{VM}}$ as obtained from the numerical simulation. The higher resolution at lower pass energies is attributed to charging of the slits, which could narrow their effective width. Note that $\Delta E/E_0 < s/r_0$ due to the enhanced dispersion caused by the space charge

exactly proportional to the pass energy $E_{0\,\mathrm{VM}}$. A possible cause for this effect will be discussed shortly. In all cases the energy dispersion $E_{0\,\mathrm{VM}}/\Delta E_{1/2\,\mathrm{VM}}$ is significantly enlarged by the effect of the space charge (cf. (4.52) for example).

The key result to be presented here with respect to the pre-monochromator is the monochromatic current produced by the device as a function of the resolution (Fig. 8.7). The output current is proportional to $\Delta E_{\mathrm{VM}}^{1.5}$, a consequence of the fact that the energy width of the feed beam is proportional to the pass energy E_0. In Fig. 8.7 we also show the calculated dependence of the monochromatic current on the energy resolution using the numerical simulation programs discussed in Chap. 5. The potentials on the top and bottom cover plates are adjusted so as to produce a nearly parallel beam in the vertical plane when the monochromator is fed with such a parallel beam, as shown for example in Fig. 5.4. This was achieved by setting the voltage at the top and bottom plate to $eU_\mathrm{D} = -0.55\,E_0$. This setting is consistent with the values for eU_D found experimentally to work best. In the simulation, we have also set the width of the energy distribution of the feed beam equal to the experimentally observed one. The maximum aperture angle $\alpha_{1\mathrm{m}}$ was set to 1.5° and $\beta_{1\mathrm{m}} = 2°$. The results of the simulation are shown in Figs. 8.6 and 8.7 as a dashed and dotted line, respectively. In both cases the agreement between experimental data and the theoretical simulation is quite close. We must keep in mind that there is no adjustable parameter in the theory. The result does not depend critically on the choice of $\beta_{1\mathrm{m}}$ as long as the latter remains small.

The remaining difference between the results of the simulation and the experiment is essentially due to the fact that the angle $\alpha_{1\mathrm{m}}$ does not remain as small as 1.5° at low pass energies. In order to have an estimate of the input angles provided by the emission system we have calculated the angular distribution and in particular the cutoff angle $\alpha_{1\mathrm{m}}$ for the slot lens cathode as a function of the

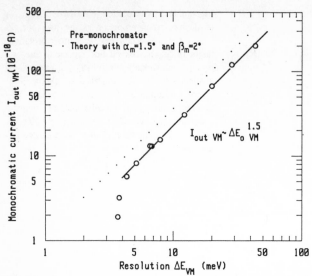

Fig. 8.7. Monochromatic current produced by the pre-monochromator vs the resolution. The dotted line is the result of the numerical simulation as described in Chap. 5. The potentials in the simulation are chosen to match the experimental data. The same holds for the energy width of the feed beam. The angular apertures α_{1m} and β_{1m} of the feed beam are not amenable to a direct experimental investigation, though they can be estimated from the simulation of the emission system. Data and theory refer to an exit aperture retardation factor of 5

potential on the entrance aperture of the pre-monochromator. In this calculation we have used the potentials actually applied to the lenses in the experiment. We have disregarded the space charge though, because our simulation of the space charge is not appropriate for the deep space charge regime in which the emission system operates. The result of the simulation is shown in Fig. 8.8. One sees clearly that the angular aperture exceeds the optimum value of 1.5° when the potential on the entrance slit drops below about 5 eV. The enlargement of α_{1m} has a significant effect on the resolution for lower pass energies and accounts at least qualitatively for the fact that the resolution levels off to a constant value at low pass energies.

We did not attempt to simulate the pre-monochromator with the angular apertures α_{1m} taken from the simulation of the emission system because the effect of enlarged angular apertures at low pass energies is intertwined with another effect which escapes the possibilities of numerical simulation. In Fig. 8.6 we notice that the experimentally obtained resolution is actually better than the theoretical resolution! The most likely reason is that the entrance and/or exit slit of the pre-monochromator are charged by the heavy current load impinging on the apertures. Negative charge on the apertures effectively narrows the slits, which results in higher resolution. The effect of charging is experimentally quite noticeable when the cathode is switched off for a while and turned on again: the monochromatic beam takes a considerable time to come back but comes back immediately if the pre-monochromator and emission system are readjusted.

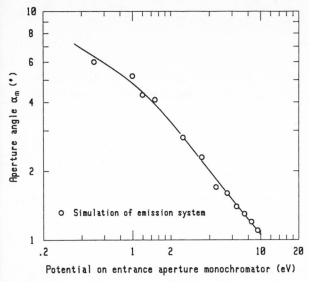

Fig. 8.8. Aperture angle α_{1m} of electrons entering the pre-monochromator with an energy equivalent to the maximum of the Maxwellian energy distribution at 2000 K. The values of α_{1m} are obtained from a numerical simulation of the slot lens emission system (Fig. 6.4) with a cathode tip size of ≈ 0.05 mm. Lens voltages are taken from experiment, apart from the repeller, which is adjusted to create a focus at the entrance aperture of the pre-monochromator. The repeller potential deviates from experiment because (i) the exact position of the cathode tip is not known in reality and (ii) because of charging. We have also determined the complete angular distribution and find the distribution to be closer to a triangular than a rectangular one. We do not know, however, how representative this result (obtained for small currents) is for the real system, which operates under heavy space charge conditions

8.4 The Second Monochromator and the Lens System

In Fig. 8.7 we have seen that a monochromator resolution of about 5 meV corresponding to about 7 meV for the entire spectrometer can be achieved with a high monochromatic current of almost 1 nA at the sample, provided that the monochromator has the extended path length. Earlier single path instruments [8.1] produced much lower monochromatic currents since the effects of space charge on the first-order focus were not properly taken into account. At the current status of electron energy loss spectroscopy, overall resolutions of 7 meV are rarely acceptable in most applications. The use of a second monochromator therefore becomes indispensable. One of the issues arising with double-stage monochromatisation is the matching of the resolution of the two monochromators. We addressed the issue earlier in Chap. 5 but there we assumed that the energy distribution of the feed beam delivered by the emission system remained constant with the pass energy of the pre-monochromators (which may be the case for some other designs). With the energy distribution of the feed beam narrowing roughly proportional to the pass energy, the output current j_{out} takes the form

$$j_{out} \sim \Delta\theta_{sc} \Delta E_{1/2}^{3/2} , \tag{8.9}$$

where $\Delta\theta_{sc}$ is the extension of the deflection angle introduced in Chap. 4. The output current of the pre-monochromator must match the optimum input current of the second monochromator, which also scales as the extension of the deflection angle $\Delta\theta_{sc}$ and as $\Delta E_{1/2}^{3/2}$. Hence, an optimum match of monochromator and pre-monochromator is achieved when

$$\Delta E_{HM} = \Delta E_{VM} \left(\frac{\Delta\theta_{sc\,HM}}{\Delta\theta_{sc\,VM}} \right)^{2/3} \tag{8.10}$$

with VM and HM referring to the first and second monochromator, respectively. For our specific design ($\Delta\theta_{sc\,HM} = 7°$; $\Delta\theta_{sc\,VM} = 29°$, $\Delta\theta_{sc}$ values in the absence of negative bias) one should therefore have

$$\Delta E_{HM} \sim 0.38 \, \Delta E_{VM} . \tag{8.11}$$

The practical use of this equation is somewhat limited by the fact that the resolution of the pre-monochromator is not a very well-defined function of the pass energy (Fig. 8.6) and the retardation factor. For the purpose of finding the optimum pass energy of the pre-monochromator in relation to the pass energy of the second monochromator, and hence the optimum retardation factor F, one determines experimentally the current at the detector and the total energy width of the spectrometer versus the retardation factor. As the retardation factor is increased, the current rises but so does the energy width $\Delta E_{1/2\,tot}$ (Fig. 8.9). An optimum is approximately given by the maximum in the performance, defined as $I_D/\Delta E_{tot}^3$ where I_D denotes the detector current. Figure 8.9 shows that this optimum retardation factor occurs at about $F = 8$. The resolution of the monochromator and pre-monochromator were then 2.9 and 8.1 meV, respectively, in good agreement with (8.10, 11). To achieve the best possible resolution of the spectrometer, it may be preferable to back off from the optimum retardation factor.

Let us pause for a moment and consider the result presented in Fig. 8.9 concerning the resolution in greater detail. According to the theory of space charge flow in monochromators, and specifically from Fig. 5.10, one would expect the energy width ΔE to pass through a minimum when the output current of the pre-monochromator matched the particular input current of the monochromator for which the first order focus occurred at the exit aperture. Experimentally the minimum is not observed when the input current is varied by means of a variation of the retardation factor. At the same time, the monochromator resolution does not agree with that calculated theoretically with the optimum entrance aperture angle α_{1m}. The reason is that the actual entrance aperture angle α_{1m} is larger than the optimum aperture angle of $\sim 3.5°$ (Fig. 5.8). From the general electron optical properties of retarding monochromators (Sect. 3.2), we know that the exit aperture angle α_{2m} and the entrance apertures angle α_{1m} of the retarding pre-monochromator are then related by (3.20),

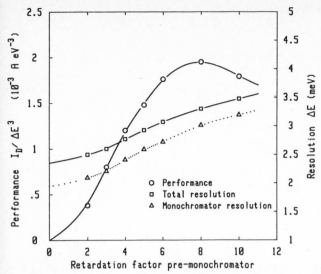

Fig. 8.9. Total resolution ΔE_{tot} of the spectrometer, resolution of the monochromator alone, and performance factor $j_D/\Delta E_{tot}^3$ vs the retardation factor of the pre-monochromator. The data refer to a pass energy of 0.214 eV of the monochromator

$$\alpha_{2m} = \alpha_{1m} F^{0.25} , \tag{8.12}$$

where F is the retardation factor. Taking the data in Fig. 8.8 and $F = 8$ the aperture angle of the second monochromator exceeds the optimum value of 3.5 when the pass energy of the monochromator drops below about 0.45 eV. The effect of an enlarged aperture angle α_{1m} on the monochromator resolution is that with lower path energy the energy width $\Delta E_{1/2}$ is no longer proportional to the pass energy. This is clearly seen in the data displayed in Fig. 8.10. By using the aperture angles obtained for the beam produced by the emission system (Fig. 8.8) and transforming them into input aperture angles of the monochromator with (8.12), the energy width $\Delta E_{1/2}$ of the beam leaving the monochromator may be estimated from (8.1). The full line in Fig. 8.10 is the result for $\beta_{1m} = 0$ (i.e. small β). The agreement between the data and the model is quite reasonable. We take the agreement as evidence that the angular aperture calculated for the emission system is not unreasonable. Further evidence for this is obtained from the transmission of the monochromator. As remarked earlier, the transmission consists of two parts, one (T_E) arising from the fact that the monochromator selects an energy band from the beam feeding the monochromator (3.48) and a second (T_α), arising from the aperture angle α_{1m} (3.30). The total transmission is the product of T_α and T_E, to a good approximation. Since we are not interested in the obvious factor T_E we have plotted

$$T_\alpha = I_{out}\sqrt{1 + (\Delta E_{in}/\Delta E_{out})^2}/I_{in} \tag{8.13}$$

in Fig. 8.11. The data refer to a retardation factor of $F = 8$. The figure shows two

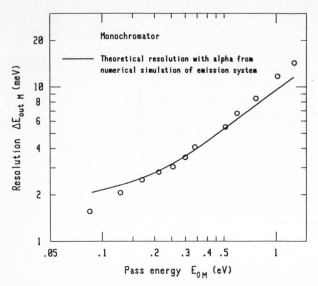

Fig. 8.10. Full width at half maximum of the energy distribution of electrons leaving the monochromator vs the pass energy. The data points are calculated from the total spectrometer resolution after eliminating the analyser resolution, which was independently determined (Fig. 8.1). The full line is a calculation, which uses the angular apertures α_{1m} of the beam produced by the emission system according to a numerical calculation

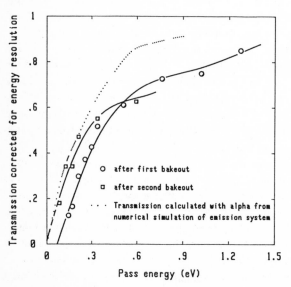

Fig. 8.11. Transmission T_α of the monochromator vs the pass energy. The dotted line is again calculated using the α_{1m} values of the emission system

sets of data, one obtained after an additional bakeout of the spectrometer in order to reduce the effect of spurious surface potentials. Clearly the second bakeout increased the performance of the spectrometer considerably at low pass energies. Nevertheless the general trend of the data to fall to zero transmission as the pass energy approaches zero is not affected. Once again, we try the hypothesis of ascribing the reduction in the transmission to the enlarged angular aperture of the feed beam. By taking the calculated aperture angles from Fig. 8.8 and by using (3.30), the transmission T_α is calculated and the result is the dotted line in Fig. 8.11. The match to the data is quite close. In particular, the drop towards zero at low pass energies is well reproduced. The discrepancy at higher pass energies may be caused either by an additional effect of the β_{1m}-aperture angle or, more likely, by larger aperture angles α_{1m} than those obtained from the simulation of the emission system.

The most relevant property of the spectrometer for practical application is the monochromatic current at the sample versus the resolution of the monochromator. The two sets of data before and after a second bakeout are shown in Fig. 8.12. The data were obtained with a retardation factor of $F = 8$. We have not found any indication of an extra broadening of the energy distribution due to the electron–electron scattering in dense beams known as the Boersch effect [8.2]. The monochromatic current delivered by the pre-monochromator is also shown in Fig. 8.12 as a dashed line. For low resolution, the current produced by the two-stage monochromatisation is higher than for a single-stage monochro-

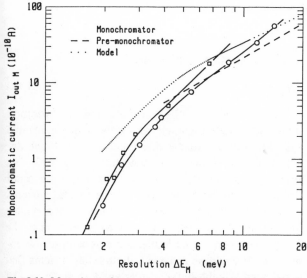

Fig. 8.12. Monochromatic current vs the width of the energy distribution (FWHM) of the electron beam leaving the monochromator. The data points are obtained in two series of measurements with a constant retarding factor of 8. The squares are data taken after a second bakeout of the spectrometer. The dashed line is the monochromatic current provided by the pre-monochromator according to the experimental data in Fig. 8.7. The dotted line is the theoretical calculation described in the text

mator, though not significantly. The second stage does permit a higher resolution to be achieved, however. If the aperture angles did not increase with lower pass energies the slope of the current versus resolution should follow the dashed line, down to the 1–2 meV range, providing an order of magnitude higher currents there. Again we compare the data in Fig. 8.12 with the theoretical results using the aperture angles of the emission system. According to (5.10) the optimum input current of the monochromator is

$$I_{in} = 0.35 \, k \, E_0^{3/2} \Delta\theta_{sc} \left(\alpha_{1m} + 0.16 \, \frac{\Delta E_{in}}{E_0} \right) . \tag{8.14}$$

The prefactor 0.35 corresponds to the case in which the compression voltage is adjusted to keep the beam parallel in the vertical plane under space charge conditions, namely, $eU_D = 0.55 \, E_0$. This potential was applied in the experiment because it led, as expected, to the highest current at the detector. For the energy part of the transmission T_E we have (3.48)

$$T_E = \left[1 + \left(\frac{\Delta E_{in}}{\Delta E_{out}} \right)^2 \right]^{-1/2} . \tag{8.15}$$

The angular part is (3.30)

$$T_\alpha = \begin{cases} 1 - (|C_{\alpha\alpha}|/3s)\alpha_{1m}^2 , & T_\alpha > \frac{2}{3} , \\ \frac{2}{3}\alpha_{1m}^{-1}(s/|C_{\alpha\alpha}|)^{1/2} , & T_\alpha < \frac{2}{3} . \end{cases} \tag{8.16}$$

For the full width at half maximum we take (3.56)

$$\Delta E_{1/2} = E_0(s/D + 0.47 \, \alpha_{1m}^2) \tag{8.17}$$

and the dispersion D is, according to (3.17) and (4.76),

$$D = 0.966 \, r_0(1 + 1.3 \, \Delta\theta_{sc}) . \tag{8.18}$$

By writing further $\Delta E_{in} = \Delta E_{out}/0.38$ from (8.12), we are now in a position to calculate the expected monochromatic current, entirely without any adjustable parameter. Hence, this is the key test for all the theoretical investigations described in Chaps. 4 and 5. The result is shown as a dotted line in Fig. 8.12. The line clearly follows the general trend of the data and is off merely by a small margin. We conclude therefore that the theory of space charge limited currents in monochromators is not only mathematically correct but has also, a matter of equal importance, considered the appropriate experimental parameters in the numerical analysis. A second conclusion is that higher currents with a low energy width in the range of 1–2 meV could be achieved, if emission systems that provided the same current and energy width but a smaller aperture angle α_{1m} of the beam could be constructed. The point is illustrated in Fig. 8.13. We assume here that the aperture angle α_{1m} scales according to

$$\alpha_{1m} = \frac{a}{\sqrt{E_0}} , \tag{8.19}$$

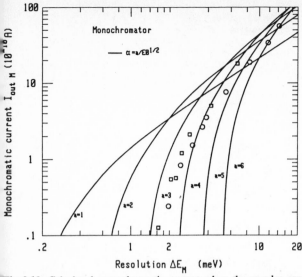

Fig. 8.13. Calculated monochromatic current when the angular aperture is $\alpha_{1m} = a/\sqrt{E_0}$ with $a = 1°$–$6°$. The experimental results obtained so far are also shown for comparison

which is, very crudely, the behaviour of the data in Fig. 8.8. The resulting currents are plotted in Fig. 8.13 for $a = 1°$–$6°$.

High resolution spectrometers with usable current levels at 1 meV resolution could thus be constructed if appropriate emission systems could be found, capable of bringing a current of about 50 nA with an energy width of about 60 meV into a slit of 0.3 mm × 2 mm with an angular aperture of 2°. Whether or not this can be achieved remains to be seen, but there is certainly room for further development [8.3].

Our final comment is on the lens system. With the particular spectrometer tested and described here we have used the movable lens with which the acceptance angle at the sample can be changed. The other lens systems described in Chap. 7 have also been tested and used in several spectrometers. They have performed according to expection. The high transmission that can be achieved with properly calculated lens systems is illustrated in Fig. 8.14. There, the current falling onto the entrance aperture of the analyser and the current at the detector are plotted versus the scattering angle. The latter is varied by moving the entire package of monochromators and one lens system around a pivot at the sample position in the scattering chamber. The current on the entrance aperture drops as the image of the exit slit of the monochromator falls into the entrance slit of the analyser. The current does not drop completely to zero because the spot is enlarged by image aberrations. The ratio of the current drop of 0.257 nA to the maximum current of 0.32 nA is the transmission of the lens, here 80%. The data refer to pass energies of 0.3 and 0.255 in monochromator and analyser, respectively, and to an impact energy of 4 eV at the sample position. The spectrometer resolution was 4.3 meV (FWHM). Together with the reduction of the

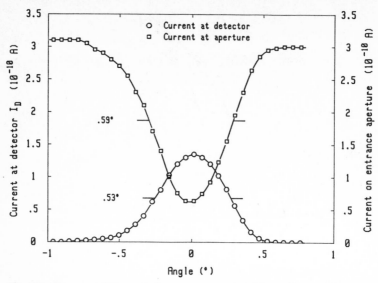

Fig. 8.14. The current falling onto the entrance aperture of the analyser and the current at the detector placed after the exit slit of the analyser vs the scattering angle at the sample position. The transmission of the lens system, the transmission of the analyser, and the angular resolution of the spectrometer may be determined from the plot

current on the entrance aperture, the current at the detector rises to a maximum value of 0.125 nA. The ratio of the current at the detector to the current entering the analyser is 0.49. This value is the transmission of the analyser including the energy resolution factor, which is about $T_E = 0.44$, so that $T_\alpha \approx 0.90$. The full width at half maximum indicated in Fig. 8.14 marks the angular resolution of the spectrometer at the target, $\alpha_t \approx 0.26°$. This value refers to the situation in which the lenses are close to monochromator and analyser as in Fig. 7.17. The value of $\alpha_t = 0.26°$ corresponds to a wave vector transfer $k_\perp$ of 4.8×10^{-3} Å^{-1} or to a transfer width of 208 Å.

High transmission of the lens can be achieved down to rather high resolution. In Fig. 8.15 we display the monochromatic current at the detector versus the resolution of the spectrometer, together with the monochromatic current emerging from the monochromator versus the monochromator resolution. The curves essentially differ by a constant factor, which is due to the product of analyser transmission and lens transmission and to the larger energy width of the entire spectrometer, arising from the convolution of the analyser and monochromator resolution functions. Since detector currents of 0.1–1 pA (in the direct beam) are sufficient for most applications in electron energy loss spectroscopy, the spectrometer can provide a resolution below 2 meV. Very recently we replaced the LaB$_6$ cathode used earlier by a tungsten cathode with an extra sharp needle tip welded onto the hairpin-shaped heating wire. This was in the hope of achieving a smaller aperture angle of the feed beam of the first monochromator and hence of improving the ultimate resolution according to Fig. 8.13. With this modifica-

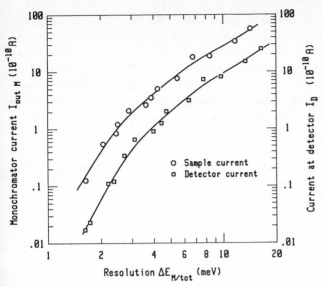

Fig. 8.15. The monochromatic current at the detector vs the total energy resolution of the spectrometer (□) and the monochromatic current emerging from the monochromator vs the monochromator resolution (o). The latter is frequently quoted in publications describing spectrometers while the former set of data is of more relevance in applications

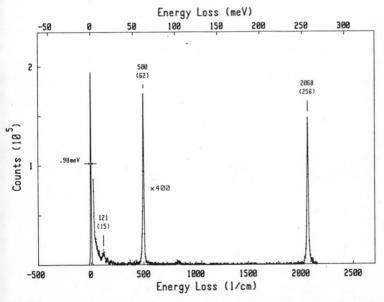

Fig. 8.16. Sample spectrum of a saturated CO layer on an Ir(100) 1×1 surface. The spectrum, with a resolution of 0.98 meV (1 meV $= 8.065$ cm^{-1}), was obtained with a modern electron energy loss spectrometer. The sampling time is 3 s per data point so that the count rate is about 7×10^4 s^{-1} in the elastic channels and up to 150 s^{-1} in the inelastic. The spectrum was recorded in 2 cm^{-1} increments

tion a resolution of $\Delta E < 1\,\mathrm{meV}$ was obtained. We note, however, that it is quite difficult with this particular cathode to find the optimum set of potentials in the emission system so that only electrons emerging from the fine tip enter the first monochromator. An example is shown in Fig. 8.16. The spectrum refers to a saturated monolayer of CO on an unreconstructed Ir(100) surface at room temperature. The resolution measured as the full width at half maximum of the elastic beam is 0.98 meV. The actual spectrometer resolution may be even better since losses due to electron-hole-pair creation tend to broaden the "elastic" signal by a few tenths of a meV. One also notices the additional broadening in the vibration losses, which is a genuine physical effect caused by dephasing and lifetime broadening. This demonstrates that the resolution of electron spectrometers is close to the physically meaningful limit.

References

Chapter 1

1.1 Examples are:
E. Yagasaki, R.I. Masel: Surf. Sci. **226**, 51 (1990)
C.T. Kao, G.S. Blackmann, M.A. v. Hove, G.A. Somorjai: Surf. Sci. **224**, 77 (1989)
S.K. So, R. Franchy, W. Ho: J. Chem. Phys. **91**, 5701 (1989)
X.Y. Zhu, S.R. Hatch, A. Campion, J.M. White: J. Chem. Phys. **91**, 5011 (1989)
Dezheng Wang, Kai Wu, Yuming Cao, Runsheng Zhai, Xiexian Guo: Surf. Sci. **223**, L927 (1989)
M.T. Paffett, S.C. Gebhard, R.G. Windham, B.E. Koel: Surf. Sci. **223**, 449 (1989)
J. Paul: Surf. Sci. **224**, 348 (1989)
1.2 Examples are:
B. Voigtländer, S. Lehwald, H. Ibach: Surf. Sci. **225**, 162 (1990)
B. Voigtländer, S. Lehwald, H. Ibach: Surf. Sci. **208**, 113 (1989)
M. Wuttig, R. Franchy, H. Ibach: Surf. Sci. **224**, L979 (1989)
W. Daum, H. Ibach, J.E. Müller: Phys. Rev. Lett. **59**, 1593 (1987)
1.3 Examples are:
T. Aizawa, R. Souda, S. Otani, Y. Ishizawa, C. Oshima: Phys. Rev. Lett. **64**, 768 (1990)
B. Voigtländer, D. Bruchmann, S. Lehwald, H. Ibach: Surf. Sci. **225**, 151 (1990)
B. Voigtländer, S. Lehwald, H. Ibach: Surf. Sci. **225**, 162 (1990)
B. Voigtländer, S. Lehwald, H. Ibach, K.P. Bohnen, K.M. Ho: Phys. Rev. B **40**, 8068 (1989)
1.4 Examples are:
J.J. Pireaux, P.A. Thiry, R. Sporken, R. Caudano: Surf. Interface Anal. **15**, 189 (1990)
J.A. Schäfer, D.J. Frankel, G.J. Lapeyre: Z. Phys. B **79**, 259 (1990)
L.H. Dubois, G.P. Schwartz: Phys. Rev. B **40**, 8336 (1989)
Z.C. Ying, W. Ho: J. Chem. Phys. **91**, 2689 (1989)
P.O. Hahn, M. Grundner, A. Schnegg, H. Jacob: Appl. Surf. Sci. **39**, 436 (1989)
Ph. Lambin, J.P. Vigneron, A.A. Lucas: Phys. Rev. B **32**, 8203 (1985)
B.N.J. Persson, J.E. Demuth: Phys. Rev. B **30**, 5968 (1984)
R. Matz, H. Lüth: Phys. Rev. Lett. **46**, 500 (1981)
1.5 Examples are:
D.N. Belton, S.J. Schmieg: J. Appl. Phys. **66**, 4223 (1989)
Ming-Cheng Wu, P.J. Møller: Phys. Rev. B **40**, 6063 (1989)
C. Tatsuyama, T. Tanbo, N. Nakayama: Appl. Surf. Sci. **41**, 539 (1989)
Y. Suda, D. Lubben, T. Motooka, J.E. Greene: J. Vac. Sci. Technol. B **7**, 1171 (1989)
R. Sporken, P.A. Thiry, P. Xhonneux, R. Caudano, J.P. Delrue: Appl. Surf. Sci. **41**, 201 (1989)
C. Mariani, M. Grazia Betti, U. del Pennino: Phys. Rev. B **40**, 8095 (1989)
L. Fotiadis, R. Kaplan: Appl. Phys. Lett. **55**, 2538 (1989)
1.6 H. Hopster, D.L. Abraham: Phys. Rev. B **40**, 7054 (1989)
J. Kirschner, D. Rebenstorff, H. Ibach: Phys. Rev. Lett. **53**, 698 (1984)

1.7   M. Rocca, H. Ibach, S. Lehwald, T.S. Rahman: In *Structure and Dynamics of Surfaces I*, ed. by W. Schommers, P. von Blanckenhagen, Topics Curr. Phys., Vol. 41 (Springer, Berlin, Heidelberg 1986) p.245
      G. Ertl, J. Küppers: *Low Energy Electrons and Surface Chemistry* (VCH, Weinheim 1985)
      J. Kirschner, D. Rebenstorff, H. Ibach: Phys. Rev. Lett. **53**, 698 (1984)
      H. Ibach, D.L. Mills: *Electron Energy Loss Spectroscopy and Surface Vibrations* (Academic, New York 1982)
      J. Fink: Recent developments in energy-loss spectroscopy. Adv. Electron. Electron Phys. **75**, 121 (1989)
1.8   C.E. Kuyatt, J.A. Simpson: Rev. Sci. Instrum. **38**, 103 (1967)
1.9   H. Foitzheim, H. Ibach, S. Lehwald: Rev. Sci. Instrum. **46**, 1325 (1975)
1.10 D. Roy, J.D. Carette: In *Electron Spectroscopy for Surface Analysis*, ed. by H. Ibach, Topics Curr. Phys., Vol. 4 (Springer, Berlin, Heidelberg 1977) p.13
1.11 B.A. Sexton: J. Vac. Sci. Technol. **16**, 1033 (1979)
1.12 L.L.Kesmodel: J. Vac. Sci. Technol. A **1**, 1456 (1983)
1.13 J.A. Stroscio, W. Ho: Rev. Sci. Instrum. **57**, 1483 (1986)
1.14 R. Franchy, H. Ibach: Surf. Sci. **155**, 15 (1985)
1.15 S.D. Kevan, L.H. Dubois: Rev. Sci. Instrum. **55**, 1604 (1984)
1.16 R.L. Strong, J.L. Erskine: Rev. Sci. Instrum. **55**, 1304 (1984)
1.17 C. Oshima, R. Franchy, H. Ibach: Rev. Sci. Instrum. **54**, 1042 (1983)
1.18 C. Oshima, R. Souda, M. Aono, Y. Ishizawa: Rev. Sci. Instrum. **56**, 227 (1986)
1.19 P.W. Hawkes, E. Kasper: *Principles of Electron Optics* (Academic, London 1989)

Chapter 2

2.1   C.E. Kuyatt, J.Y. Simpson: Rev. Sci. Instrum. **38**, 103 (1967)
2.2   K. Hünlich: Diplomarbeit, Aachen (1984), unpublished
2.3   E.W. Schmid, G. Spitz, W. Lösch: *Theoretische Physik mit dem Personal Computer* (Springer, Berlin, Heidelberg 1987), p.101
      Ximen Jiye: *Aberration Theory in Electron and Ion Optics* (Academic, Orlando, FL 1986), p.368
2.4   K.R. Spangenberg: *Vacuum Tubes* (McGraw-Hill, New York 1948), p.170
2.5   H. Ibach, D.L. Mills: *Electron Energy Loss Spectroscopy and Surface Vibrations* (Academic, New York 1982)

Chapter 3

3.1   D. Roy, J.D. Carette: Appl. Phys. Lett. **16**, 413 (1970); Can. J. Phys. **49**, 2138 (1971)
3.2   H. Wollnik, H. Ewald: Nuclear Instrum. Methods **36**, 93 (1965)
3.3   O. Klemperer: *Electron Optics* (Cambridge University Press, Cambridge 1953), pp.136 ff.
3.4   P. Bryce, R.L. Dalghlish, J.C. Kelly: Can. J. Phys. **51**, 574 (1973)
3.5   C. Oshima, R. Franchy, H. Ibach: Rev. Sci. Instrum. **54**, 1042 (1983)
3.6   C. Oshima, R. Souda, M. Aono, Y. Ishizawa: Rev. Sci. Instrum. **56**, 227 (1986)
3.7   R. Herzog: Z. Phys. **97**, 556 (1935); ibid. **41**, 18 (1940)
3.8   H. Froitzheim, H. Ibach, S. Lehwald: Rev. Sci. Instrum. **46**, 1325 (1975)
3.9   L.L. Kesmodel: J. Vac. Sci. Technol. A **1**, 1456 (1983)
3.10 L.L. Kesmodel: US Patent 4,559,449 (1985)

3.11 H. Froitzheim: J. Electron Spectroscopy Relat. Phenom. **34**, 11 (1984)
3.12 K. Jost: J. Phys. Sci. Instrum. **12**, 1006 (1979)
3.13 S.D. Kevan, L.H. Dubois: Rev. Sci. Instrum. **55**, 1604 (1984)

Chapter 4

4.1 E.G. Johnson, A.O. Nier: Phys. Rev. **91**, 10 (1953)
4.2 H. Ibach, D.L. Mills: *Electron Energy Loss Spectroscopy and Surface Vibrations* (Academic, New York 1982), p.46
4.3 H.A. Engelhardt, W. Bäck, D. Menzel, H. Liebl: Rev. Sci. Instrum. **52**, 835 (1981)
      H.A. Engelhardt, A. Zartner, D. Menzel: Rev. Sci. Instrum. **52**, 1161 (1981)

Chapter 6

6.1 O. Klemperer: *Electron Optics* (Cambridge University Press, Cambridge 1953)
6.2 C. Kunz: In *Synchrotron Radiation*, ed. by C. Kunz, Topics Curr. Phys., Vol. 10 (Springer, Berlin, Heidelberg 1979), pp.1ff.
6.3 K.R. Spangenberg: *Vacuum Tubes* (McGraw-Hill, New York 1948)
6.4 W. Knauer: Optik **54**, 211 (1979)
6.5 A.B. El-Kareh, J.C.J. El-Kareh: *Electron Beams, Lenses and Optics* (Academic, New York 1970)
6.6 P. Grivet: *Electron Optics* (Pergamon, Oxford 1972)
6.7 E.W. Schmid, G. Spitz, W. Lösch: *Theoretische Physik mit dem Personal Computer* (Springer, Berlin, Heidelberg 1987)
6.8 H. Ehrhardt, L. Langhans, F. Linder, H.S. Taylor: Phys. Rev. **173**, 222 (1968)

Chapter 7

7.1 A.A. Lucas, M. Sunjić: Prog. Surf. Sci. **2**, 75 (1972); Phys. Rev. Lett. **26**, 229 (1971)
      M. Sunjić, A.A. Lucas: Phys. Rev. B **3**, 719 (1971)
7.2 A.A. Lucas, J.P. Vigneron: Solid State Commun. **49**, 327 (1984)
7.3 For a review of electron spectroscopy in transmission see: H. Raether: *Solid State Excitations by Electrons*, Springer Tracts in Modern Physics, Vol. 38 (Springer, Berlin, Heidelberg 1965) p.84
      J. Daniels, C.v. Festenberg, H. Raether, K. Zeppenfeld: Springer Tracts in Modern Physics, Vol. 54 (Springer, Berlin, Heidelberg 1970), p.77
7.4 H. Froitzheim, H. Ibach, D.L. Mills: Phys. Rev. B **11**, 4980 (1974)
7.5 H. Ibach, D.L. Mills: *Electron Energy Loss Spectroscopy and Surface Vibrations* (Academic, New York 1982)
7.6 H. Ibach: In 9th Intl. Vacuum Cong. and 5th Intl. Cong. on Surface Science, Madrid, ed. by F.L. de Segovia (Asociacion española del vacio y sus applicaciones, Madrid 1983)
7.7 S.Y. Tong, C.H. Li, D.L. Mills: Phys. Rev. Lett. **44**, 407 (1980)
      C.H. Li, S.Y. Tong, D.L. Mills: Phys. Rev. B **21**, 3057 (1980)
      S.Y. Tong, C.H. Li, D.L. Mills: Phys. Rev. B **24**, 806 (1981)
7.8 Patents: EP 0013 003, DE 28 51 743, US 4 845 361

7.9 Lens systems designed with 2D-computer codes have been described in: R.L. Strong, J.L. Erskine: Rev. Sci. Instrum. **55**, 1304 (1984); A. Sellidj, J.L. Erskine: Rev. Sci. Instrum. **61**, 49 (1990)

7.10 Mu-Liang Xu, B.M. Hall, S.Y. Tong, M. Rocca, H. Ibach, S. Lehwald, J.E. Black: Phys. Rev. Lett. **54**, 1171 (1985)

7.11 M. Born, E. Wolf: *Principles of Optics* (Pergamon, Oxford 1980)

7.12 H. Froitzheim, H. Ibach, S. Lehwald: Rev. Sci. Instrum. **46**, 1325 (1975)

Chapter 8

8.1 H. Ibach, D.L. Mills: *Electron Energy Loss Spectroscopy and Surface Vibrations* (Academic, New York 1982) pp. 49 ff.

8.2 P.W. Hawkes, E. Kasper: *Principles of Electron Optics II* (Academic, London 1989) pp. 1004 ff.

8.3 While this volume was in print a significant improvement was achieved for the current at very resolution by using a different emission system. The data for the monochromatic current track the curve with $a = 2$ in Fig. 8.13

Subject Index